The Meaning of It All

The Meaning of It All

*Thoughts of a
Citizen Scientist*

Richard P. Feynman

HELIX BOOKS

PERSEUS BOOKS
Reading, Massachusetts

We are very grateful to Carl Feynman and Michelle Feynman
for making this book possible.
We are also grateful to Dr. Judith Goodstein and her staff
at the Archives of the California Institute of Technology, without
whose kind hospitality and cooperation these lectures might have
remained hidden for all time!

Many of the designations used by manufacturers and sellers to distinguish their products are claimed as trademarks. Where those designations appear in this book and Perseus Books was aware of a trademark claim, the designations have been printed in initial capital letters.

Library of Congress Cataloging-in-Publication Data
Feynman, Richard Phillips.
 The meaning of it all: thoughts of a citizen scientist/Richard P. Feynman.
  p. cm.
 Includes index.
 ISBN 0-201-36080-2
 1. Science—Social aspects—Miscellanea. 2. Science and religion.
I. Title.
Q 175.5.F52 1998
500—dc21   97-48250
                                                               CIP

Perseus Books is a member of the Perseus Books Group

Jacket design by Suzanne Heiser
Jacket photogrph courtesy of the Archives, California Institute of Technology
Text design by Irving Perkins Associates
Set in 11.5 Simoncini Garamond by Pagesetters, inc.

3 4 5 6 7 8 9 10-DOH-0201009998
Third Printing, May 1998

Find Helix Books on the World Wide Web at
http://www.aw.com/gb/

Contents

THE UNIVERSITY OF WASHINGTON
proudly presents the second

John Danz Lecturer

PROFESSOR RICHARD P. FEYNMAN
Physicist, CALIFORNIA INSTITUTE OF TECHNOLOGY

in a series of three closely related lectures

"A SCIENTIST LOOKS AT SOCIETY"

topics will include
"THIS UNSCIENTIFIC AGE"
"SCIENCE AND HUMAN VALUES"
"SCIENCE AND MAN'S FUTURE"

in the series, Dr. Feynman explores problems in the borderline between
science and philosophy, religion, and society

complimentary

8 p.m.	8 p.m.	8 p.m.
April 23	April 25	April 27
Meany Hall	Health Sciences Auditorium	Health Sciences Auditorium

Publisher's Note

It is our great honor to share these brilliant and illuminating lectures, published here for the first time.

In April 1963, Richard P. Feynman was invited to give a three-night series of lectures at the University of Washington (Seattle) as part of the John Danz Lecture Series. Here is Feynman the man revealing, as only he could, his musings on society, on the conflict between science and religion, on peace and war, on our universal fascination with flying saucers, on faith healing and telepathy, on people's distrust of politicians—indeed on all the concerns of the modern citizen-scientist.

Pure gold, pure poetry, pure Feynman.

I

The Uncertainty of Science

I WANT TO ADDRESS myself directly to the impact of science on man's ideas in other fields, a subject Mr. John Danz particularly wanted to be discussed. In the first of these lectures I will talk about the nature of science and emphasize particularly the existence of doubt and uncertainty. In the second lecture I will discuss the impact of scientific views on political questions, in particular the question of national enemies, and on religious questions. And in the third lecture I will describe how society looks to me—I could say how society looks to a scientific man, but it is only how it looks to me—and what future scientific discoveries may produce in terms of social problems.

What do I know of religion and politics? Several friends in the physics departments here and in other places laughed and said, "I'd like to come and hear what you have to say. I never knew you were interested very much in those things." They mean, of course, I am interested, but I would not dare to talk about them.

In talking about the impact of ideas in one field on ideas in another field, one is always apt to make a fool of oneself. In these days of specialization there are too few people who have such a deep understanding of two departments of our knowledge that they do not make fools of themselves in one or the other.

The ideas I wish to describe are old ideas. There is

practically nothing that I am going to say tonight that could not easily have been said by philosophers of the seventeenth century. Why repeat all this? Because there are new generations born every day. Because there are great ideas developed in the history of man, and these ideas do not last unless they are passed purposely and clearly from generation to generation.

Many old ideas have become such common knowledge that it is not necessary to talk about or explain them again. But the ideas associated with the problems of the development of science, as far as I can see by looking around me, are not of the kind that everyone appreciates. It is true that a large number of people do appreciate them. And in a university particularly most people appreciate them, and you may be the wrong audience for me.

New in this difficult business of talking about the impact of the ideas of one field on those of another, I shall start at the end that I know. I do know about science. I know its ideas and its methods, its attitudes toward knowledge, the sources of its progress, its mental discipline. And therefore, in this first lecture, I shall talk about the science that I know, and I shall leave the more ridiculous of my statements for the next two lectures, at which, I assume, the general law is that the audiences will be smaller.

What is science? The word is usually used to mean one of three things, or a mixture of them. I do not think we need to be precise—it is not always a good idea to be too precise. Science means, sometimes, a special method of finding things out. Sometimes it means the body of knowledge arising from the things found out. It may also mean the new things you can do when you have found something out, or the actual doing of new things. This last field is usually called technology—but if you look at the science section in *Time* magazine you will find it covers about 50 percent what new things are found out and about 50 percent what new things can be and are being done. And so the popular definition of science is partly technology, too.

I want to discuss these three aspects of science in reverse order. I will begin with the new things that you can do—that is, with technology. The most obvious characteristic of science is its application, the fact that as a consequence of science one has a power to do things. And the effect this power has had need hardly be mentioned. The whole industrial revolution would almost have been impossible without the development of science. The possibilities today of producing quantities of food adequate for such a large population, of controlling sickness—the very fact that there can be free men without the necessity of slavery for full production—are very likely the result of the development of scientific means of production.

Now this power to do things carries with it no instructions on how to use it, whether to use it for good or for evil. The product of this power is either good or evil, depending on how it is used. We like improved production, but we have problems with automation. We are happy with the development of medicine, and then we worry about the number of births and the fact that no one dies from the diseases we have eliminated. Or else, with the same knowledge of bacteria, we have hidden laboratories in which men are working as hard as they can to develop bacteria for which no one else will be able to find a cure. We are happy with the development of air transportation and are impressed by the great airplanes, but we are aware also of the severe horrors of air war. We are pleased by the ability to communicate between nations, and then we worry about the fact that we can be snooped upon so easily. We are excited by the fact that space can now be entered; well, we will undoubtedly have a difficulty there, too. The most famous of all these imbalances is the development of nuclear energy and its obvious problems.

Is science of any value?

I think a power to do something is of value. Whether the result is a good thing or a bad thing depends on how it is used, but the power is a value.

Once in Hawaii I was taken to see a Buddhist temple. In the temple a man said, "I am going to tell you something that you will never forget." And then he said, "To every man is given the key to the gates of heaven. The same key opens the gates of hell."

And so it is with science. In a way it is a key to the

gates of heaven, and the same key opens the gates of hell, and we do not have any instructions as to which is which gate. Shall we throw away the key and never have a way to enter the gates of heaven? Or shall we struggle with the problem of which is the best way to use the key? That is, of course, a very serious question, but I think that we cannot deny the value of the key to the gates of heaven.

All the major problems of the relations between society and science lie in this same area. When the scientist is told that he must be more responsible for his effects on society, it is the applications of science that are referred to. If you work to develop nuclear energy you must realize also that it can be used harmfully. Therefore, you would expect that, in a discussion of this kind by a scientist, this would be the most important topic. But I will not talk about it further. I think that to say these are scientific problems is an exaggeration. They are far more humanitarian problems. The fact that how to work the power is clear, but how to control it is not, is something not so scientific and is not something that the scientist knows so much about.

Let me illustrate why I do not want to talk about this. Some time ago, in about 1949 or 1950, I went to Brazil to teach physics. There was a Point Four program in those days, which was very exciting—everyone was going to help the underdeveloped countries. What they needed, of course, was technical know-how.

In Brazil I lived in the city of Rio. In Rio there are

hills on which are homes made with broken pieces of wood from old signs and so forth. The people are extremely poor. They have no sewers and no water. In order to get water they carry old gasoline cans on their heads down the hills. They go to a place where a new building is being built, because there they have water for mixing cement. The people fill their cans with water and carry them up the hills. And later you see the water dripping down the hill in dirty sewage. It is a pitiful thing.

Right next to these hills are the exciting buildings of the Copacabana beach, beautiful apartments, and so on.

And I said to my friends in the Point Four program, "Is this a problem of technical know-how? They don't know how to put a pipe up the hill? They don't know how to put a pipe to the top of the hill so that the people can at least walk uphill with the empty cans and downhill with the full cans?"

So it is not a problem of technical know-how. Certainly not, because in the neighboring apartment buildings there are pipes, and there are pumps. We realize that now. Now we think it is a problem of economic assistance, and we do not know whether that really works or not. And the question of how much it costs to put a pipe and a pump to the top of each of the hills is not one that seems worth discussing, to me.

Although we do not know how to solve the problem, I would like to point out that we tried two things, technical know-how and economic assistance. We are

The Uncertainty of Science

discouraged with them both, and we are trying something else. As you will see later, I find this encouraging. I think that to keep trying new solutions is the way to do everything.

Those, then are the practical aspects of science, the new things that you can do. They are so obvious that we do not need to speak about them further.

The next aspect of science is its contents, the things that have been found out. This is the yield. This is the gold. This is the excitement, the pay you get for all the disciplined thinking and hard work. The work is not done for the sake of an application. It is done for the excitement of what is found out. Perhaps most of you know this. But to those of you who do not know it, it is almost impossible for me to convey in a lecture this important aspect, this exciting part, the real reason for science. And without understanding this you miss the whole point. You cannot understand science and its relation to anything else unless you understand and appreciate the great adventure of our time. You do not live in your time unless you understand that this is a tremendous adventure and a wild and exciting thing.

Do you think it is dull? It isn't. It is most difficult to convey, but perhaps I can give some idea of it. Let me start anywhere, with any idea.

For instance, the ancients believed that the earth was the back of an elephant that stood on a tortoise that swam in a bottomless sea. Of course, what held up the sea was another question. They did not know the answer.

The belief of the ancients was the result of imagination. It was a poetic and beautiful idea. Look at the way we see it today. Is that a dull idea? The world is a spinning ball, and people are held on it on all sides, some of them upside down. And we turn like a spit in front of a great fire. We whirl around the sun. That is more romantic, more exciting. And what holds us? The force of gravitation, which is not only a thing of the earth but is the thing that makes the earth round in the first place, holds the sun together and keeps us running around the sun in our perpetual attempt to stay away. This gravity holds its sway not only on the stars but between the stars; it holds them in the great galaxies for miles and miles in all directions.

This universe has been described by many, but it just goes on, with its edge as unknown as the bottom of the bottomless sea of the other idea—just as mysterious, just as awe-inspiring, and just as incomplete as the poetic pictures that came before.

But see that the imagination of nature is far, far greater than the imagination of man. No one who did not have some inkling of this through observations could ever have imagined such a marvel as nature is.

Or the earth and time. Have you read anywhere, by any poet, anything about time that compares with real time, with the long, slow process of evolution? Nay,

The Uncertainty of Science

I went too quickly. First, there was the earth without anything alive on it. For billions of years this ball was spinning with its sunsets and its waves and the sea and the noises, and there was no thing alive to appreciate it. Can you conceive, can you appreciate or fit into your ideas what can be the meaning of a world without a living thing on it? We are so used to looking at the world from the point of view of living things that we cannot understand what it means not to be alive, and yet most of the time the world had nothing alive on it. And in most places in the universe today there probably is nothing alive.

Or life itself. The internal machinery of life, the chemistry of the parts, is something beautiful. And it turns out that all life is interconnected with all other life. There is a part of chlorophyll, an important chemical in the oxygen processes in plants, that has a kind of square pattern; it is a rather pretty ring called a benzine ring. And far removed from the plants are animals like ourselves, and in our oxygen-containing systems, in the blood, the hemoglobin, there are the same interesting and peculiar square rings. There is iron in the center of them instead of magnesium, so they are not green but red, but they are the same rings.

The proteins of bacteria and the proteins of humans are the same. In fact it has recently been found that the protein-making machinery in the bacteria can be given orders from material from the red cells to produce red cell proteins. So close is life to life. The universality of the deep chemistry of living things is indeed a fantastic and beautiful thing. And all the time we human beings have been too proud even to recognize our kinship with the animals.

Or there are the atoms. Beautiful—mile upon mile of one ball after another ball in some repeating pattern in a crystal. Things that look quiet and still, like a glass of water with a covered top that has been sitting for several days, are active all the time; the atoms are leaving the surface, bouncing around inside, and coming back. What looks still to our crude eyes is a wild and dynamic dance.

And, again, it has been discovered that all the world is made of the same atoms, that the stars are of the same stuff as ourselves. It then becomes a question of where our stuff came from. Not just where did life come from, or where did the earth come from, but where did the stuff of life and of the earth come from? It looks as if it was belched from some exploding star, much as some of the stars are exploding now. So this piece of dirt waits four and a half billion years and evolves and changes, and now a strange creature stands here with instruments and talks to the strange creatures in the audience. What a wonderful world!

Or take the physiology of human beings. It makes no difference what I talk about. If you look closely enough at anything, you will see that there is nothing

are involved in the entire universe. And so he got, by looking at every feature of the candle, into combustion, chemistry, etc. But the introduction of the book, in describing Faraday's life and some of his discoveries, explained that he had discovered that the amount of electricity necessary to do performic electrolysis of chemical substances is proportional to the number of atoms which are separated divided by the valence. It further explained that the principles he discovered are used today in chrome plating and the anodic coloring of aluminum, as well as in dozens of other industrial applications. I do not like that statement. Here is what Faraday said about his own discovery: "The atoms of matter are in some ways endowed or associated with electrical powers, to which they owe their most striking qualities, amongst them their mutual chemical affinity." He had discovered that the thing that determined how the atoms went together, the thing that determined the combinations of iron and oxygen which make iron oxide is that some of them are electrically plus and some of them are electrically minus, and they attract each other in definite proportions. He also discovered that electricity comes in units, in atoms. Both were important discoveries, but most exciting was that this was one of the most dramatic moments in the history of science, one of those rare moments when two great fields come together and are unified. He suddenly found that two apparently different things were different aspects of the same thing. Electricity was being studied, and chemistry was being studied. Suddenly they were two aspects of the same thing—chemical changes with the results of electrical forces. And they are still understood that way. So to say merely that the principles are used in chrome plating is inexcusable.

And the newspapers, as you know, have a standard line for every discovery made in physiology today: "The discoverer said that the discovery may have uses in the cure of cancer." But they cannot explain the value of the thing itself.

Trying to understand the way nature works involves a most terrible test of human reasoning ability. It involves subtle trickery, beautiful tightropes of logic on which one has to walk in order not to make a mistake in predicting what will happen. The quantum mechanical and the relativity ideas are examples of this.

The third aspect of my subject is that of science as a method of finding things out. This method is based on the principle that observation is the judge of whether something is so or not. All other aspects and characteristics of science can be understood directly when we understand that observation is the ultimate and final judge of the truth of an idea. But "prove" used in this way really means "test," in the same way that a hundred-proof alcohol is a test of the alcohol, and for people today the idea really should be translated as, "The exception *tests* the rule." Or, put another way, "The exception proves that the rule is wrong." That is the principle of science. If there is an exception to any rule, and if it can be proved by observation, that rule is wrong.

The exceptions to any rule are most interesting in themselves, for they show us that the old rule is wrong. And it is most exciting, then, to find out what the right rule, if any, is. The exception is studied, along with other conditions that produce similar effects. The scientist tries to find more exceptions and to determine the characteristics of the exceptions, a process that is continually exciting as it develops. He does not try to avoid showing that the rules are wrong; there is progress and excitement in the exact opposite. He tries to prove himself wrong as quickly as possible.

The principle that observation is the judge imposes a severe limitation to the kind of questions that can be answered. They are limited to questions that you can put this way: "if I do this, what will happen?" There are ways to try it and see. Questions like, "should I do this?" and "what is the value of this?" are not of the same kind.

But if a thing is not scientific, if it cannot be subjected to the test of observation, this does not mean that it is dead, or wrong, or stupid. We are not trying to argue that science is somehow good and other things are somehow not good. Scientists take all those things that *can* be analyzed by observation, and thus the things called science are found out. But there are some things left out, for which the method does not work. This does not mean that those things are unimportant. They are, in fact, in many ways the most important. In any decision for action, when you have to make up your mind what to do, there is always a "should" involved, and this cannot be worked out from "if I do this, what will happen?" alone. You say, "Sure, you see what will happen, and then you decide whether you want it to happen or not." But that is the step the scientist cannot take. You can figure out what is going to happen, but then you have to decide whether you like it that way or not.

There are in science a number of technical consequences that follow from the principle of observation as judge. For example, the observation cannot be rough. You have to be very careful. There may have been a piece of dirt in the apparatus that made the color change; it was not what you thought. You have to check the observations very carefully, and then recheck them, to be sure that you understand what all the conditions are and that you did not misinterpret what you did.

It is interesting that this thoroughness, which is a virtue, is often misunderstood. When someone says a thing has been done scientifically, often all he means is that it has been done thoroughly. I have heard people talk of the "scientific" extermination of the Jews in Germany. There was nothing scientific about it. It was only thorough. There was no question of making observations and then checking them in order to determine something. In that sense, there were "scientific" exterminations of people in Roman times and in other periods when science was not so far developed as it is today and not much attention was paid to observation. In such cases, people should say "thorough" or "thoroughgoing," instead of "scientific."

There are a number of special techniques associated with the game of making observations, and much of what is called the philosophy of science is concerned with a discussion of these techniques. The interpretation of a result is an example. To take a trivial instance, there is a famous joke about a man who complains to a friend of a mysterious phenomenon. The white horses on his farm eat more than the black horses. He worries about this and cannot understand it, until his friend suggests that maybe he has more white horses than black ones.

It sounds ridiculous, but think how many times similar mistakes are made in judgments of various kinds. You say, "My sister had a cold, and in two weeks . . ." It is one of those cases, if you think about it, in which there were more white horses. Scientific reasoning requires a certain discipline, and we should try to teach this discipline, because even on the lowest level such errors are unnecessary today.

Another important characteristic of science is its objectivity. It is necessary to look at the results of observation objectively, because you, the experimenter, might like one result better than another. You perform the experiment several times, and because of irregularities, like pieces of dirt falling in, the result varies from time to time. You do not have everything under control. You like the result to be a certain way, so the times it comes out that way, you say, "See, it comes out this particular way." The next time you do the experiment it comes out different. Maybe there was a piece of dirt in it the first time, but you ignore it.

These things seem obvious, but people do not pay enough attention to them in deciding scientific questions or questions on the periphery of science. There could be a certain amount of sense, for example, in the way you analyze the question of whether stocks went up or down because of what the President said or did not say.

Another very important technical point is that the more specific a rule is, the more interesting it is. The more definite the statement, the more interesting it is to test. If someone were to propose that the planets go around the sun because all planet matter has a kind of tendency for movement, a kind of motility, let us call it an "oomph," this theory could explain a number of other phenomena as well. So this is a good theory, is it not? No. It is nowhere near as good as a proposition that the planets move around the sun under the influence of a central force which varies exactly inversely as the square of the distance from the center. The second theory is better because it is so specific; it is so obviously unlikely to be the result of chance. It is so definite that the barest error in the movement can show that it is wrong; but the planets could wobble all over the place, and, according to the first theory, you could say, "Well, that is the funny behavior of the 'oomph.'"

So the more specific the rule, the more powerful it is, the more liable it is to exceptions, and the more interesting and valuable it is to check.

Words can be meaningless. If they are used in such a way that no sharp conclusions can be drawn, as in my example of "oomph," then the proposition they state is almost meaningless, because you can explain almost anything by the assertion that things have a tendency to motility. A great deal has been made of this by philosophers, who say that words must be defined extremely precisely. Actually, I disagree somewhat with this; I think that extreme precision of definition is often not worthwhile, and sometimes it is not possible—in fact mostly it is not possible, but I will not get into that argument here.

Most of what many philosophers say about science is really on the technical aspects involved in trying to make sure the method works pretty well. Whether these technical points would be useful in a field in which observation is not the judge I have no idea. I am not going to say that everything has to be done the same way when a method of testing different from observation is used. In a different field perhaps it is not so important to be careful of the meaning of words or that the rules be specific, and so on. I do not know.

In all of this I have left out something very important. I said that observation is the judge of the truth of an idea. But where does the idea come from? The rapid progress and development of science requires that human beings invent something to test.

It was thought in the Middle Ages that people simply make many observations, and the observations themselves suggest the laws. But it does not work that way. It takes much more imagination than that. So the next thing we have to talk about is where the new ideas come from. Actually, it does not make any difference, as long as they come. We have a way of checking whether an idea is correct or not that has nothing to do with where it came from. We simply test it against observation. So in science we are not interested in where an idea comes from.

There is no authority who decides what is a good idea. We have lost the need to go to an authority to find out whether an idea is true or not. We can read an authority and let him suggest something; we can try it out and find out if it is true or not. If it is not true, so much the worse— so the "authorities" lose some of their "authority."

The relations among scientists were at first very argumentative, as they are among most people. This was true in the early days of physics, for example. But in physics today the relations are extremely good. A scientific argument is likely to involve a great deal of laughter and uncertainty on both sides, with both sides thinking up experiments and offering to bet on the outcome. In physics there are so many accumulated observations that it is almost impossible to think of a new idea which is different from all the ideas that have been thought of before and yet that agrees with all the observations that have already been made. And so if you get anything new from anyone, anywhere, you welcome it, and you do not argue about why the other person says it is so.

Many sciences have not developed this far, and the situation is the way it was in the early days of physics, when there was a lot of arguing because there were not so many observations. I bring this up because it is interesting that human relationships, if there is an independent way of judging truth, can become unargumentative.

Most people find it surprising that in science there is no interest in the background of the author of an idea or in his motive in expounding it. You listen, and if it sounds like a thing worth trying, a thing that could be tried, is different, and is not obviously contrary to something observed before, it gets exciting and worthwhile. You do not have to worry about how long he has studied or why he wants you to listen to him. In that sense it makes no difference where the ideas come from. Their real origin is unknown; we call it the imagination of the human brain, the creative imagination—it is known; it is just one of those "oomphs."

It is surprising that people do not believe that

The Uncertainty of Science

there is imagination in science. It is a very interesting kind of imagination, unlike that of the artist. The great difficulty is in trying to imagine something that you have never seen, that is consistent in every detail with what has already been seen, and that is different from what has been thought of; furthermore, it must be definite and not a vague proposition. That is indeed difficult.

Incidentally, the fact that there are rules at all to be checked is a kind of miracle; that it is possible to find a rule, like the inverse square law of gravitation, is some sort of miracle. It is not understood at all, but it leads to the possibility of prediction—that means it tells you what you would expect to happen in an experiment you have not yet done.

It is interesting, and absolutely essential, that the various rules of science be mutually consistent. Since the observations are all the same observations, one rule cannot give one prediction and another rule another prediction. Thus, science is not a specialist business; it is completely universal. I talked about the atoms in physiology; I talked about the atoms in astronomy, electricity, chemistry. They are universal; they must be mutually consistent. You cannot just start off with a new thing that cannot be made of atoms.

It is interesting that reason works in guessing at the rules, and the rules, at least in physics, become reduced. I gave an example of the beautiful reduction of the rules in chemistry and electricity into one rule, but there are many more examples.

The rules that describe nature seem to be mathematical. This is not a result of the fact that observation is the judge, and it is not a characteristic necessity of science that it be mathematical. It just turns out that you can state mathematical laws, in physics at least, which work to make powerful predictions. Why nature is mathematical is, again, a mystery.

I come now to an important point. The old laws may be wrong. How can an observation be incorrect? If it has been carefully checked, how can it be wrong? Why are physicists always having to change the laws? The answer is, first, that the laws are not the observations and, second, that experiments are always inaccurate. The laws are guessed laws, extrapolations, not something that the observations insist upon. They are just good guesses that have gone through the sieve so far. And it turns out later that the sieve now has smaller holes than the sieves that were used before, and this time the law is caught. So the laws are guessed; they are extrapolations into the unknown. You do not know what is going to happen, so you take a guess.

For example, it was believed—it was discovered—that motion does not affect the weight of a thing—that if you spin a top and weigh it, and then weigh it when it has stopped, it weighs the same. That is the result of an observation. But you cannot weigh something to the infinitesimal number of decimal places, parts in a billion. But we now understand that a spinning top weighs more than a top which is not spinning by a few parts in less than a billion. If the top spins fast enough so that the speed of the edges approaches 186,000 miles a second, the weight increase is appreciable—but not until then. The first experiments were performed with tops that spun at speeds much lower than 186,000 miles a second. It seemed then that the mass of the top spinning and not spinning was exactly the same, and someone made a guess that the mass never changes.

How foolish! What a fool! It is only a guessed law, an extrapolation. Why did he do something so unscientific? There was nothing unscientific about it; it was only uncertain. It would have been unscientific *not* to guess. It has to be done because the extrapolations are the only things that have any real value. It is only the principle of what you think will happen in a case you have not tried that is worth knowing about. Knowledge is of no real value if all you can tell me is what happened yesterday. It is necessary to tell what will happen tomorrow if you do something—not necessary, but fun. Only you must be willing to stick your neck out.

Every scientific law, every scientific principle, every statement of the results of an observation is some kind of a summary which leaves out details, because nothing can be stated precisely. The man simply forgot—he should have stated the law "The mass doesn't change *much* when the speed isn't *too high*." The game is to make a specific rule and then see if it will go through the sieve. So the specific guess was that the mass never changes at all. Exciting possibility! It does no harm that it turned out not to be the case. It was only uncertain, and there is no harm in being uncertain. It is better to say something and not be sure than not to say anything at all.

It is necessary and true that all of the things we say in science, all of the conclusions, are uncertain, because they are only conclusions. They are guesses as to what is going to happen, and you cannot know what will happen, because you have not made the most complete experiments.

It is curious that the effect on the mass of a spinning top is so small you may say, "Oh, it doesn't make any difference." But to get a law that is right, or at least one that keeps going through the successive sieves, that goes on for many more observations, requires a tremendous intelligence and imagination and a complete revamping of our philosophy, our understanding of space and time. I am referring to the relativity theory. It turns out that the tiny effects that turn up always require the most revolutionary modifications of ideas.

Scientists, therefore, are used to dealing with doubt and uncertainty. All scientific knowledge is uncertain. This experience with doubt and uncertainty is important. I believe that it is of very great value, and one that extends beyond the sciences. I believe that to solve any problem that has never been solved before, you have to leave the door to the unknown ajar. You have to permit the possibility that you do not have it exactly right. Otherwise, if you have made up your mind already, you might not solve it.

When the scientist tells you he does not know the answer, he is an ignorant man. When he tells you he has a hunch about how it is going to work, he is uncertain about it. When he is pretty sure of how it is going to work, and he tells you, "This is the way it's going to work, I'll bet," he still is in some doubt. And it is of paramount importance, in order to make progress, that we recognize this ignorance and this doubt. Because we have the doubt, we then propose looking in new directions for new ideas. The rate of the development of science is not the rate at which you make observations alone but, much more important, the rate at which you create new things to test.

If we were not able or did not desire to look in any new direction, if we did not have a doubt or recognize ignorance, we would not get any new ideas. There would be nothing worth checking, because we would know what is true. So what we call scientific knowledge today is a body of statements of varying degrees of certainty. Some of them are most unsure; some of them are nearly sure; but none is absolutely certain. Scientists are used to this. We know that it is consistent to be able to live and not know. Some people say, "How can you *live* without knowing?" I do not know what they mean. I always live without knowing. That is easy. How you get to know is what I want to know.

This freedom to doubt is an important matter in the sciences and, I believe, in other fields. It was born of a struggle. It was a struggle to be permitted to doubt, to be unsure. And I do not want us to forget the importance of the struggle and, by default, to let the thing fall away. I feel a responsibility as a scientist who knows the great value of a satisfactory philosophy of ignorance, and the progress made possible by such a philosophy, progress which is the fruit of freedom of thought. I feel a responsibility to proclaim the value of this freedom and to teach that doubt is not to be feared, but that it is to be welcomed as the possibility of a new potential for human beings. If you know that you are not sure, you have a chance to improve the situation. I want to demand this freedom for future generations.

Doubt is clearly a value in the sciences. Whether it is in other fields is an open question and an uncertain matter. I expect in the next lectures to discuss that very point and to try to demonstrate that it is important to doubt and that doubt is not a fearful thing, but a thing of very great value.

II

The Uncertainty of Values

W<small>E ARE ALL SAD</small> when we think of the wondrous potentialities that human beings seem to have and when we contrast these potentialities with the small accomplishments that we have. Again and again people have thought that we could do much better. People in the past had, in the nightmare of their times, dreams for the future, and we of their future have, although many of those dreams have been surpassed, to a large extent the same dreams. The hopes for the future today are in a great measure the same as they were in the past. At some time people thought that the potential that people had was not developed because everyone was ignorant and that education was the solution to the problem, that if all people were educated, we could perhaps all be Voltaires. But it turns out that falsehood and evil can be taught as easily as good. Education is a great power, but it can work either way. I have heard it said that the communication between nations should lead to an understanding and thus a solution to the problem of developing the potentialities of man. But the means of communication can be channeled and choked. What is communicated can be lies as well as truth, propaganda as well as real and valuable information. Communication is a strong force, also, but either for good or evil. The applied sciences, for a while, were thought to free men of material difficulties at least, and there is some good in the record, especially, for example, in medicine. On the other hand, scientists are working now in secret laboratories to develop the diseases that they were so careful to control.

Everybody dislikes war. Today our dream is that peace will be the solution. Without the expense of armaments, we can do whatever we want. And peace is a great force for good or for evil. How will it be for evil? I do not know. We will see, if we ever get peace. We have, clearly, peace as a great force, as well as material power, communication, education, honesty, and the ideals of many dreamers. We have more forces of this kind to control today than did the ancients. And maybe we are doing it a little bit better than most of them could do. But what we ought to be able to do seems gigantic compared to our confused accomplishments. Why is this? Why can't we conquer ourselves? Because we find that even the greatest forces and abilities don't seem to carry with them any clear instructions on how to use them. As an example, the great accumulation of understanding as to how the physical world behaves only convinces one that this behavior has a kind of meaninglessness about it. The sciences do not directly teach good and bad.

Throughout all the ages, men have been trying to fathom the meaning of life. They realize that if some direction or some meaning could be given to the whole thing, to our actions, then great human forces would be unleashed. So, very many answers have been given to the question of the meaning of it all. But they have all been of different sorts. And the proponents of one idea have looked with horror at the actions of the believers of another—horror because from a disagreeing point of view all the great potentialities of the race were being channeled into a false and confining blind alley. In fact, it is from the history of the enormous monstrosities that have been created by false belief that philosophers have come to realize the fantastic potentialities and wondrous capacities of human beings.

The dream is to find the open channel. What, then, is the meaning of it all? What can we say today to dispel the mystery of existence? If we take everything into account, not only what the ancients knew, but also all those things that we have found out up to today that they didn't know, then I think that we must frankly admit that we do not know. But I think that in admitting this we have probably found the open channel.

Admitting that we do not know and maintaining perpetually the attitude that we do not know the direction *necessarily* to go permit a possibility of alteration, of thinking, of new contributions and new discoveries for the problem of developing a way to do what we want ultimately, even when we do not know what we want.

Looking back at the worst times, it always seems that they were times in which there were people who believed with absolute faith and absolute dogmatism in something. And they were so serious in this matter that they insisted that the rest of the world agree with them. And then they would do things that were directly inconsistent with their own beliefs in order to maintain that what they said was true.

So I have developed in a previous talk, and I want to maintain here, that it is in the admission of ignorance and the admission of uncertainty that there is a hope for the continuous motion of human beings in some direction that doesn't get confined, permanently blocked, as it has so many times before in various periods in the history of man. I say that we do not know what is the meaning of life and what are the right moral values, that we have no way to choose them and so on. No discussion can be made of moral values, of the meaning of life and so on, without coming to the great source of systems of morality and descriptions of meaning, which is in the field of religion.

And so I don't feel that I could give three lectures on the subject of the impact of scientific ideas on other ideas without frankly and completely discussing the relation of science and religion. I don't know why I should even have to start to make an excuse for doing this, so I won't continue to try to make such an excuse. But I would like to begin a discussion of the question of a conflict, if any, between science and religion. I described more or less what I meant by science, and I have to tell you what I mean by religion, which is extremely difficult, because different people mean different things. But in the discussion that I want to talk about here I mean the everyday, ordinary, church-going kind of religion, not the elegant theology that belongs to it, but the way ordinary people believe, in a more or less conventional way, about their religious beliefs.

I do believe that there is a conflict between science and religion, religion more or less defined that way. And in order to bring the question to a position that is easy to discuss, by making the thing very definite, instead of trying to make a very difficult theological study, I would present a problem which I see happens from time to time.

A young man of a religious family goes to the university, say, and studies science. As a consequence of his study of science, he begins, naturally, to doubt as it is necessary in his studies. So first he begins to doubt, and then he begins to disbelieve, perhaps, in his father's God. By "God" I mean the kind of personal God, to which one prays, who has something to do with creation, as one prays for moral values, perhaps. This phenomenon happens often. It is not an isolated or an imaginary case. In fact, I believe, although I have no direct statistics, that more than half of the scientists do not believe in their father's God, or in God in a conventional sense. Most scientists do not believe in it. Why? What happens? By answering this question I think that we will point up most clearly the problems of the relation of religion and science.

Well, why is it? There are three possibilities. The first is that the young man is taught by the scientists, and I have already pointed out, they are atheists, and so their evil is spread from the teacher to the student, perpetually. . . . Thank you for the laughter. If you take this point of view, I believe it shows that you know less of science than I know of religion.

The second possibility is to suggest that because a little knowledge is dangerous, that the young man just learning a little science thinks he knows it all, and to suggest that when he becomes a little more mature he will understand better all these things. But I don't think so. I think that there are many mature scientists, or men who consider themselves mature—and if you didn't know about their religious beliefs ahead of time you would decide that they are mature—who do not believe in God. As a matter of fact, I think that the answer is the exact reverse. It isn't that he knows it all, but he suddenly realizes that he doesn't know it all.

The third possibility of explanation of the phenomenon is that the young man perhaps doesn't understand science correctly, that science cannot disprove God, and that a belief in science and religion is consistent. I agree that science cannot disprove the existence of God. I absolutely agree. I also agree that a belief in science and religion is consistent. I know many scientists who believe in God. It is not my purpose to disprove anything. There are very many scientists who do believe in God, in a conventional way too, perhaps, I do not know exactly how they believe in God. But their belief in God and their action in science is thoroughly consistent. It is consistent, but it is difficult. And what I would like to discuss here is why it is hard to attain this consistency and perhaps whether it is worthwhile to attempt to attain the consistency.

There are two sources of difficulty that the young man we are imagining would have, I think, when he studies science. The first is that he learns to doubt, that it is necessary to doubt, that it is valuable to doubt. So, he begins to question everything. The question that might have been before, "Is there a God or isn't there a God" changes to the question "How sure am I that there is a God?" He now has a new and subtle problem that is different than it was before. He has to determine how sure he is, where on the scale between absolute certainty and absolute certainty on the other side he can put his belief, because he knows that he has to have his knowledge in an unsure condition and he cannot be absolutely certain anymore. He has to make up his mind. Is it 50-50 or is it 97 percent? This sounds like a very small difference, but it is an extremely important and subtle difference. Of course it is true that the man does not usually start by doubting directly the existence of God. He usually starts by doubting some other details of the belief, such as the belief in an afterlife, or some of the details of Christ's life, or something like this. But in order to make this question as sharp as possible, to be frank with it, I will simplify it and will come right directly to the question of this problem about whether there is a God or not.

The result of this self-study or thinking, or whatever it is, often ends with a conclusion that is very close to certainty that there is a God. And it often ends, on the other hand, with the claim that it is almost certainly wrong to believe that there is a God.

Now the second difficulty that the student has when he studies science, and which is, in a measure, a kind of conflict between science and religion, because it is a human difficulty that happens when you are educated two ways. Although we may argue theologically and on a high-class philosophical level that there is no conflict, it is still true that the young man who comes from a religious family gets into some argument with himself and his friends when he studies science, so there is some kind of a conflict.

Well, the second origin of a type of conflict is associated with the facts, or, more carefully, the partial facts that he learns in the science. For example, he learns about the size of the universe. The size of the universe is very impressive, with us on a tiny particle that whirls around the sun. That's one sun among a hundred thousand million suns in this galaxy, itself among a billion galaxies. And again, he learns about the close biological relationship of man to the animals and of one form of life to another and that man is a latecomer in a long and vast, evolving drama. Can the rest be just a scaffolding for His creation? And yet again there are the atoms, of which all appears to be constructed following immutable laws. Nothing can escape it. The stars are made of the same stuff, the animals are made of the same stuff—but in some such complexity as to mysteriously appear alive.

It is a great adventure to contemplate the universe, beyond man, to contemplate what it would be like without man, as it was in a great part of its long history and as it is in a great majority of places. When this objective view is finally attained, and the mystery and majesty of matter are fully appreciated, to then turn the objective eye back on man viewed as matter, to view life as part of this universal mystery of greatest depth, is to sense an experience which is very rare, and very exciting. It usually ends in laughter and a delight in the futility of trying to understand what this atom in the universe is, this thing—atoms with curiosity—that looks at itself and wonders why it wonders. Well, these scientific views end in awe and mystery, lost at the edge in uncertainty, but they appear to be so deep and so impressive that the theory that it is all arranged as a stage for God to watch man's struggle for good and evil seems inadequate.

Some will tell me that I have just described a religious experience. Very well, you may call it what you will. Then, in that language I would say that the young man's religious experience is of such a kind that he finds the religion of his church inadequate to describe, to encompass that kind of experience. The God of the church isn't big enough.

Perhaps. Everyone has different opinions.

Suppose, however, our student does come to the view that individual prayer is not heard. I am not trying to disprove the existence of God. I am only trying to give you some understanding of the origin of the difficulties that people have who are educated from two different points of view. It is not possible to disprove the existence of God, as far as I know. But is true that it is difficult to take two different points of view that come from different directions. So let us suppose that this particular student is particularly difficult and does come to the conclusion that individual prayer is not heard. Then what happens? Then the doubting machinery, his doubts, are turned on ethical problems. Because, as he was educated, his religious views had it that the ethical and moral values were the word of God. Now if God maybe isn't there, maybe the ethical and moral values are wrong. And what is very interesting is that they have survived almost intact. There may have been a period when a few of the moral views and the ethical positions of his religion seemed wrong, he had to think about them, and many of them he returned to.

But my atheistic scientific colleagues, which does not include all scientists—I cannot tell by their behavior, because of course I am on the same side, that they are particularly different from the religious ones, and it seems that their moral feelings and their understandings of other people and their humanity and so on apply to the believers as well as the disbelievers. It seems to me that there is a kind of independence between the ethical and moral views and the theory of the machinery of the universe.

Science makes, indeed, an impact on many ideas associated with religion, but I do not believe it affects, in any very strong way, the moral conduct and ethical views. Religion has many aspects. It answers all kinds of questions. I would, however, like to emphasize three aspects.

The first is that it tells what things are and where they came from and what man is and what God is and what properties God has and so on. I'd like, for the purposes of this discussion, to call those the *metaphysical* aspects of religion.

And then it says how to behave. I don't mean in the terms of ceremonies or rituals or things like that, but I mean how to behave in general, in a moral way. This we could call the *ethical* aspect of religion.

And finally, people are weak. It takes more than the right conscience to produce right behavior. And even though you may feel you know what you are supposed to do, you all know that you don't do things the way you would like yourself to do them. And one of the powerful aspects of religion is its inspirational aspects. Religion gives inspiration to act well. Not only that, it gives inspiration to the arts and to many other activities of human beings.

Now these three aspects of religion are very closely interconnected, in the religion's view. First of all, it usually goes something like this: that the moral values are the word of God. Being the word of God connects the ethical and metaphysical aspects of religion. And finally, that also inspires the inspiration, because if you are working for God and obeying God's will, you are in some way connected to the universe, your actions have a meaning in the greater world, and that is an inspiring aspect. So these three aspects are very well integrated and interconnected. The difficulty is that science occasionally conflicts with the first two categories, that is with the ethical and with the metaphysical aspects of religion.

There was a big struggle when it was discovered that the earth rotates on its axis and goes around the sun. It was not supposed to be the case according to the religion of the time. There was a terrible argument and the outcome was, in that case, that religion retreated from the position that the earth stood at the center of the universe. But at the end of the retreat there was no change in the moral viewpoint of the religion. There was another tremendous argument when it was found likely that man descended from the animals. Most religions have retreated once again from the metaphysical position that it wasn't true. The result is no particular change in the moral view. You see that the earth moves around the sun, yes, then does that tell us whether it is or is not good to turn the other cheek? It is this conflict associated with these metaphysical aspects that is doubly difficult because the facts conflict. Not only the facts, but the spirits conflict. Not only are there difficulties about whether the sun does or doesn't rotate around the earth, but the spirit or attitude toward the facts is also different in religion from what it is in science. The uncertainty that is necessary in order to appreciate nature is not easily correlated with the feeling of certainty in faith, which is usually associated with deep religious belief. I do not believe that the scientist can have that same certainty of faith that very deeply religious people have. Perhaps they can. I don't know. I think that it is difficult. But anyhow it seems that the metaphysical aspects of religion have nothing to do with the ethical values, that the moral values seem somehow to be outside of the scientific realm. All these conflicts don't seem to affect the ethical value.

I just said that ethical values lie outside the scientific realm. I have to defend that, because many people think the other way. They think that scientifically we should get some conclusions about moral values.

I have several reasons for that. You see, if you don't have a good reason, you have to have several reasons, so I have four reasons to think that moral values lie outside the scientific realm. First, in the past there were conflicts. The metaphysical positions have changed, and there have been practically no effects on the ethical views. So there must be a hint that there is an independence.

Second, I already pointed out that, I think at least, there are good men who practice Christian ethics and don't believe in the divinity of Christ. Incidentally, I forgot to say earlier that I take a provincial view of religion. I know that there are many people here who have religions that are not Western religions. But in a subject as broad as this it is better to take a special example, and you have to just translate to see how it looks if you are an Arab or a Buddhist, or whatever.

The third thing is that, as far as I know in the gathering of scientific evidence, there doesn't seem to be anywhere, anything that says whether the Golden Rule is a good one or not. I don't have any evidence of it on the basis of scientific study.

And finally I would like to make a little philosophical argument—this I'm not very good at, but I would like to make a little philosophical argument to explain why theoretically I think that science and moral questions are independent. The common human problem, the big question, always is "Should I do this?" It is a question of action. "What should I do? Should I do this?" And how can we answer such a question? We can divide it into two parts. We can say, "If I do this what will happen?" That doesn't tell me whether I should do this. We still have another part, which is "Well, do I want that to happen?" In other words, the first ques-

tion—"If I do this what will happen?"—is at least sus-
ceptible to scientific investigation; in fact, it is a typical
scientific question. It doesn't mean we know what will
happen. Far from it. We never know what is going to
happen. The science is very rudimentary. But, at least it
is in the realm of science we have a method to deal with
it. The method is "Try it and see"—we talked about
that—and accumulate the information and so on. And
so the question "If I do it what will happen?" is a typ-
ically scientific question. But the question "Do I want
this to happen"—in the ultimate moment—is not. Well,
you say, if I do this, I see that everybody is killed, and,
of course, I don't want that. Well, how do you know
you don't want people killed? You see, at the end you
must have some ultimate judgment.

You could take a different example. You could say,
for instance, "If I follow this economic policy, I see
there is going to be a depression, and, of course, I don't
want a depression." Wait. You see, only knowing that
it is a depression doesn't tell you that you do not want
it. You have then to judge whether the feelings of power
you would get from this, whether the importance of the
country moving in this direction is better than the cost
to the people who are suffering. Or maybe there would
be some sufferers and not others. And so there must at
the end be some ultimate judgment somewhere along
the line as to what is valuable, whether people are valuable, whether life is valuable. Deep in the end—you may

follow the argument of what will happen further and further along—but ultimately you have to decide "Yeah, I want that" or "No, I don't." And the judgment there is of a different nature. I do not see how by knowing what will happen alone it is possible to know if ultimately you want the last of the things. I believe, therefore, that it is impossible to decide moral questions by the scientific technique, and that the two things are independent.

Now the inspirational aspect, the third aspect of religion, is what I would like to turn to, and that brings me to a central question that I would like to ask you all, because I have no idea of the answer. The source of inspiration today, the source of strength and comfort in any religion, is closely knit with the metaphysical aspects. That is, the inspiration comes from working for God, from obeying His will, and so on. Now an emotional tie expressed in this manner, the strong feeling that you are doing right, is weakened when the slightest amount of doubt is expressed as to the existence of God. So when a belief in God is uncertain, this particular method of obtaining inspiration fails. I don't know the answer to this problem, the problem of maintaining the real value of religion as a source of strength and of courage to most men while at the same time not requiring an absolute faith in the metaphysical system. You may think that it might be possible to invent a metaphysical system for religion which will state things in such a way that science will never find itself in disagreement. But I do not think that it is possible to take an adventurous and ever-expanding science that is going into an unknown, and to tell the answer to questions ahead of time and not expect that sooner or later, no matter what you do, you will find that some answers of this kind are wrong. So I do not think that it is possible to not get into a conflict if you require an absolute faith in metaphysical aspects, and at the same time I don't understand how to maintain the real value of religion for inspiration if we have some doubt as to that. That's a serious problem.

Western civilization, it seems to me, stands by two great heritages. One is the scientific spirit of adventure—the adventure into the unknown, an unknown that must be recognized as unknown in order to be explored, the demand that the unanswerable mysteries of the universe remain unanswered, the attitude that all is uncertain. To summarize it: humility of the intellect.

The other great heritage is Christian ethics—the basis of action on love, the brotherhood of all men, the value of the individual, the humility of the spirit. These two heritages are logically, thoroughly consistent. But logic is not all. One needs one's heart to follow an idea. If people are going back to religion, what are they going back to? Is the modern church a place to give comfort to a man who doubts God? More, one who disbelieves in God? Is the modern church the place to give comfort and encouragement to the value of such doubts? So far, haven't we drawn strength and comfort to maintain the one or the other of these consistent heritages in a way which attacks the values of the other? Is this unavoidable? How can we draw inspiration to support these two pillars of Western civilization so that they may stand together in full vigor, mutually unafraid? That, I don't know. But that, I think, is the best I can do on the relationship of science and religion, the religion which has been in the past and still is, therefore, a source of moral code as well as inspiration to follow that code.

Today we find, as always, a conflict between nations, in particular a conflict between the two great sides, Russia and the United States. I insist that we are uncertain of our moral views. Different people have different ideas of what is right and wrong. If we are uncertain of our ideas of what is right and wrong, how can we choose in this conflict? Where is the conflict? With economic capitalism versus government control of economics, is it absolutely clear and perfectly important which side is right? We must remain uncertain. We may be pretty sure that capitalism is better than government control, but we have our own government controls. We have 52 percent; that is the corporate income tax control.

There are arguments between religion on the one hand, usually meant to represent our country, and atheism on the other hand, supposed to represent the Russians. Two points of view—they are only two points of view—no way to decide. There is a problem of human values, or the value of the state, the question of how to deal with crimes against the state—different points of view—we can only be uncertain. Do we have a real conflict? There is perhaps some progress of dictatorial government toward the confusion of democracy and the confusion of democracy toward somewhat more dictatorial government. Uncertainty apparently means no conflict. How nice. But I don't believe it. I think there is a definite conflict. I think that Russia represents danger in saying that the solution to human problems is known, that all effort should be for the state, for that means there is no novelty. The human machine is not allowed to develop its potentialities, its surprises, its varieties, its new solutions for difficult problems, its new points of view.

The government of the United States was developed under the idea that nobody knew how to make a government, or how to govern. The result is to invent a system to govern when you don't know how. And the way to arrange it is to permit a system, like we have, wherein new ideas can be developed and tried out and thrown away. The writers of the Constitution knew of the value of doubt. In the age that they lived, for instance, science had already developed far enough to show the possibilities and potentialities that are the result of having uncertainty, the value of having the openness of possibility. The fact that you are not sure means that it is possible that there is another way some day. That openness of possibility is an opportunity. Doubt and discussion are essential to progress. The United States government, in that respect, is new, it's modern, and it is scientific. It is all messed up, too. Senators sell their votes for a dam in their state and discussions get all excited and lobbying replaces the minority's chance to represent itself, and so forth. The government of the United States is not very good, but it, with the possible exception the government of England, is the greatest government on the earth today, is the most satisfactory, the most modern, but not very good.

Russia is a backward country. Oh, it is technologically advanced. I described the difference between what I like to call the science and technology. It does not apparently seem, unfortunately, that engineering and technological development are not consistent with suppressed new opinion. It appears, at least in the days of Hitler, where no new science was developed, nevertheless rockets were made, and rockets also can be made in Russia. I am sorry to hear that, but it is true that technological development, the applications of science, can go on without the freedom. Russia is backward because it has not learned that there is a limit to government power. The great discovery of the Anglo-Saxons is—they are not the only people who thought of it, but, to take the later history of the long struggle of the idea—that there can be a limit to government power. There is no free criticism of ideas in Russia. You say, "Yes, they discuss anti-Stalinism." Only in a definite form. Only to a definite extent. We should take advantage of this. Why don't we discuss anti-Stalinism too? Why don't we point out all the troubles we had with that gentleman? Why don't we point out the dangers that there are in a government that can have such a thing grow inside itself? Why don't we point out the analogies between the Stalinism that is being criticized inside of Russia and the behavior that is going on at the very same moment inside Russia? Well, all right, all right . . .

Now, I get excited, see. . . . It's only emotion. I shouldn't do that, because we should do this more scientifically. I won't convince you very well unless I make believe that it is a completely rational, unprejudiced scientific argument.

I only have a little experience in those countries. I visited Poland, and I found something interesting. The Polish people, of course, are freedom-loving people, and they are under the influence of the Russians. They can't publish what they want, but at the time when I was there, which was a year ago, they could say what they wanted, strangely enough, but not publish anything. And so we would have very lively discussions in public places on all sides of various questions. The most striking thing to remember about Poland, by the way, is that they have had an experience with Germany which is so deep and so frightening and so horrible that they cannot possibly forget it. And, therefore, all of their attitudes in foreign affairs have to do with a fear of the resurgence of Germany. And I thought while I was there of the terrible crime that would be the result of a policy on the part of the free countries which would permit once again the development of that kind of a thing in that country. Therefore, they accept Russia. Therefore, they explained to me, you see, the Russians definitely are holding down the East Germans. There is no way that the East Germans are going to have any Nazis. And there is no question that the Russians can control them. And so at least there is that buffer. And the thing that struck me as odd was that they didn't realize that one country can protect another country, and guarantee it, without dominating it completely, without living there.

The other thing they told me was very often, different individuals would call me aside and say that we would be surprised to find that, if Poland did get free of Russia and had their own government and were free, they would go along more or less the way they are going. I said, "What do you mean? I am surprised. You mean you wouldn't have freedom of speech." "Oh, no, we would have all the freedoms. We would love the freedoms, but we would have nationalized industries and so on. We believe in the socialistic ideas." I was surprised because I don't understand the problem that way. I don't think of the problem as between socialism and capitalism but rather between suppression of ideas and free ideas. If it is that free ideas and socialism are better than communism, it will work its way through. And it will be better for everybody. And if capitalism is better than socialism, it will work its way through. We have got 52 percent . . . well . . .

The fact that Russia is not free is clear to everyone, and the consequences in the sciences are quite obvious. One of the best examples is Lysenko, who has a theory of genetics, which is that acquired characteristics can be passed on to the offspring. This is probably true. The great majority, however, of genetic influences are undoubtedly of a different kind, and they are carried by the germ plasm. There are undoubtedly a few examples, a few small examples already known, in which some kind of a characteristic is carried to the next generation by direct, what we like to call *psychoplasmic*, inheritance. But the main point is that the major part of genetic behavior is in a different manner than Lysenko thinks. So he has spoiled Russia. The great Mendel, who discovered the laws of genetics, and the beginnings of the science, is dead. Only in the Western countries can it be continued, because they are not free in Russia to analyze these things. They have to discuss and argue against us all the time. And the result is interesting. Not only in this case has it stopped the science of biology, which, by the way, is the most active, most exciting, and most rapidly developing science today in the West. In Russia it is doing nothing. At the same time you would think that from an economic standpoint such a thing is impossible. But nevertheless by having the incorrect theories of inheritance and genetics, the biology of the agriculture of Russia is behind. They don't develop the hybrid corn right. They don't know how to develop better brands of potatoes. They used to know. They had the greatest potato tuber collections and so on in Russia before Lysenko than anywhere in the world. But today they have nothing of this kind. They only argue with the West.

In physics there was a time when there was trouble. In recent times there has been a great freedom for the physicist. Not a hundred percent freedom; there are different schools of thought which argue with each other. They were all in a meeting in Poland. And the Polish Intourist, the analogue of Intourist in Poland, which is call *Polorbis*, arranged a trip. And of course, there was only a limited number of rooms, and they made the mistake of putting Russians in the same room. They came down and they screamed, "For seventeen years I have never talked to that man, and I will not be in the same room with him."

There are two schools of physics. And there are the good guys and the bad guys, and it's perfectly obvious, and it's very interesting. And there are great physicists in Russia, but physics is developing much more rapidly in the West, and although it looked for a while like something good would happen there, it hasn't.

Now this doesn't mean that technology is not developing or that they are in some way backward that way, but I'm trying to show that in a country of this kind the development of ideas is doomed.

You have read about the recent phenomenon in modern art. When I was in Poland there was modern art hung in little corners in back streets. And there was the beginning of modern art in Russia. I don't know what the value of modern art is. I mean either way. But Mr. Khrushchev visited such a place, and Mr. Khrushchev decided that it looked as if this painting were painted by the tail of a jackass. My comment is, he should know.

To make the thing still more real I give you the example of a Mr. Nakhrosov who traveled in the United States and in Italy and went home and wrote what he saw. He was castigated for, I quote the castigator, "A 50-50 approach, for bourgeois objectivism." Is this a scientific country? Where did we ever get the idea that the Russians were, in some sense, scientific? Because in the early days of their revolution they had different ideas than they have now? But it is not scientific to not adopt a 50-50 approach—that is, to not understand what there is in the world in order to modify things; that is, to be blind in order to maintain ignorance.

I cannot help going on with this criticism of Mr. Nakhrosov and to tell you more about it. It was made by a man whose name is Padgovney, who is the first secretary of the Ukranian Communist Party. He said, "You told us here . . . (He was at a meeting at which the other man had just spoken, but nobody knows what he said, because it wasn't published. But the criticism was published.) You told us here you would only write the truth, the great truth, the real truth, for which you fought in the trenches of Stalingrad. That would be fine. We all advise you to write that way. (I hope he does.) Your speech, and the ideas you continue to support smack of petty bourgeois anarchy. This the party and people cannot and will not tolerate. You, Comrade Nakhrosov, had better think this over very seriously." How can the poor man think it over seriously? How can anyone think seriously about being a petty bourgeois anarchist? Can you picture an old anarchist who is a bourgeois also? And at the same time petty? The whole thing is absurd. Therefore, I hope that we can all maintain laughter and ridicule for the people like Mr. Padgovney, and at the same time try to communicate in some way to Mr. Nakhrosov that we admire and respect his courage, because we are here only at the very beginning of time for the human race. There are thousands of years in the past, and there is an unknown amount of time in the future. There are all kinds of opportunities, and there are all kinds of dangers. Man has been stopped before by stopping his ideas. Man has been jammed for long periods of time. We will not tolerate this. I hope for freedom for future generations—freedom to doubt, to develop, to continue the adventure of finding out new ways of doing things, of solving problems.

Why do we grapple with problems? We are only in

the beginning. We have plenty of time to solve the problems. The only way that we will make a mistake is that in the impetuous youth of humanity we will decide we know the answer. This is it. No one else can think of anything else. And we will jam. We will confine man to the limited imagination of today's human beings.

We are not so smart. We are dumb. We are ignorant. We must maintain an open channel. I believe in limited government. I believe that government should be limited in many ways, and what I am going to emphasize is only an intellectual thing. I don't want to talk about everything at the same time. Let's take a small piece, an intellectual thing.

No government has the right to decide on the truth of scientific principles, nor to prescribe in any way the character of the questions investigated. Neither may a government determine the aesthetic value of artistic creations, nor limit the forms of literary or artistic expression. Nor should it pronounce on the validity of economic, historic, religious, or philosophical doctrines. Instead it has a duty to its citizens to maintain the freedom, to let those citizens contribute to the further adventure and the development of the human race. Thank you.

III

This Unscientific Age

I WAS HAPPY, WHEN I got the invitation to give the John Danz Lectures, to hear that there would be three lectures, as I had thought about these ideas at great length and wanted an opportunity not to express myself in only one lecture, but to develop the ideas slowly and carefully in three lectures. I found out that I developed them slowly and carefully, completely, in two.

I have completely run out of organized ideas, but I have a large number of uncomfortable feelings about the world which I haven't been able to put into some obvious, logical, and sensible form. So, since I already contracted to give three lectures, the only thing I can do is to give this potpourri of uncomfortable feelings without having them very well organized.

Perhaps someday, when I find a real deep reason behind them all, I will be able to give them in one sensible lecture instead of this thing. Also, in case you are beginning to believe that some of the things I said before are true because I am a scientist and according to the brochure that you get I won some awards and so forth, instead of your looking at the ideas themselves and judging them directly—in other words, you see, you have some feeling toward authority—I will get rid of that tonight. I dedicate this lecture to showing what ridiculous conclusions and rare statements such a man as myself can make. I wish, therefore, to destroy any image of authority that has previously been generated.

You see, a Saturday night is a night for entertainment, and that is . . . I think I have got the right spirit now and we can go on. It is always a good to entitle a lecture in a way that nobody can believe. It is either peculiar or it is just the opposite of what you would expect. And that is the reason, of course, for calling it "This Unscientific Age." Of course if you mean by scientific the applications of technology, there is no doubt that this is a scientific age. There is no doubt at all that today we have all kinds of scientific applications which are causing us all kinds of trouble as well as giving us all kinds of advantages. And so in that sense it certainly is a scientific age. If you mean by a scientific age an age in which science is developing rapidly and advancing fully as fast as it can, then this is definitely a scientific age.

The speed at which science has been developing for the last two hundred years has been ever increasing, and we reach a culmination of speed now. We are in particular in the biological sciences, on the threshold of the most remarkable discoveries. What they are going to be I am unable to tell you. Naturally, that is the excitement of it. And the excitement that comes from turning one stone over after another and finding underneath new discoveries has been going on now perpetually for several hundred years, and it is an ever-rising crescendo. This is, in that sense, definitely a scientific age. It has been called a heroic age, by a scientist, of course. Nobody else knows about it. Sometime when history looks back at this age they will see that it was a most dramatic and remarkable age, the transformation from not knowing much about the world to knowing a great deal more than was known before. But if you mean that this is an age of science in the sense that in art, in literature, and in people's attitudes and understandings, and so forth science plays a large part, I don't think it is a scientific age at all. You see, if you take, the heroic age of the Greeks, say, there were poems about the military heroes. In the religious period of the Middle Ages, art was related directly to religion, and people's attitudes toward life were definitely closely knit to the religious viewpoints. It was a religious age. This is not a scientific age from that point of view.

Now, that there are unscientific things is not my grief. That's a nice word. I mean, that is not what I am worrying about, that there are unscientific things. That something is unscientific is not bad; there is nothing the matter with it. It is just unscientific. And *scientific* is limited, of course, to those things that we can tell about by trial and error. For example, there is the absurdity of the young these days chanting things about purple people eaters and hound dogs, something that we cannot criticize at all if we belong to the old flat foot floogie and a floy floy or the music goes down and around. Sons of mothers who sang about "come, Josephine, in my flying machine," which sounds just about as modern as "I'd like to get you on a slow boat to China." So in life, in gaiety, in emotion, in human pleasures and pursuits, and in literature and so on, there is no need to be scientific, there is no reason to be scientific. One must relax and enjoy life. That is not the criticism. That is not the point.

But if you do stop to think about it for a while, you will find that there are numerous, mostly trivial things which are unscientific, unnecessarily. For instance, there are extra seats in the front here, even though there are people [standing in the back].

While I was talking to some of the students in one of the classes, one man asked me a question, which was, "Are there any attitudes or experiences that you have when working in scientific information which you think might be useful in working with other information?"

(By the way, I will at the end say how much of the world today is sensible, rational, and scientific. It's a great deal. So, I am only taking the bad parts first. It's more fun. Then we soften it at the end. And I latched onto that as a nice organizing way to make my discussion of all the things that I think are unscientific in the world.)

I would like, therefore, to discuss some of the little tricks of the trade in trying to judge an idea. We have the advantage that we can ultimately refer the idea to experiment in the sciences, which may not be possible in other fields. But nevertheless, some of the ways of judging things, some of the experiences undoubtedly are useful in other ways. So, I start with a few examples.

The first one has to do with whether a man knows what he is talking about, whether what he says has some basis or not. And my trick that I use is very easy. If you ask him intelligent questions—that is, penetrating, interested, honest, frank, direct questions on the subject, and no trick questions—then he quickly gets stuck. It is like a child asking naive questions. If you ask naive but relevant questions, then almost immediately the person doesn't know the answer, if he is an honest man. It is important to appreciate that. And I think that I can illustrate one unscientific aspect of the world which would be probably very much better if it were more scientific. It has to do with politics. Suppose two politicians are running for president, and one goes through the farm section and is asked, "What are you going to do about the farm question?" And he knows right away—bang, bang, bang. Now he goes to the next campaigner who comes through. "What are you going to do about the farm problem?" "Well, I don't know. I used to be a general, and I don't know anything about farming. But it seems to me it must be a very difficult problem, because for twelve, fifteen, twenty years people have been struggling with it, and people say that they know how to solve the farm problem. And it must be a hard problem. So the way that I intend to solve the farm problem is to gather around me a lot of people who know something about it, to look at all the experience that we have had with this problem before, to take a certain amount of time at it,

tainty. I would like to remind you that you can be pretty sure of things even though you are uncertain, that you don't have to be so in-the-middle, in fact not at all in-the-middle. People say to me, "Well, how can you teach your children what is right and wrong if you don't know?" Because I'm pretty sure of what's right and wrong. I'm not absolutely sure; some experiences may change my mind. But I know what I would expect to teach them. But, of course, a child won't learn what you teach him.

I would like to mention a somewhat technical idea, but it's the way, you see, we have to understand how to handle uncertainty. How does something move from being almost certainly false to being almost certainly true? How does experience change? How do you handle the changes of your certainty with experience? And it's rather complicated, technically, but I'll give a rather simple, idealized example.

You have, we suppose, two theories about the way something is going to happen, which I will call "Theory A" and "Theory B." Now it gets complicated. Theory A and Theory B. Before you make any observations, for some reason or other, that is, your past experiences and other observations and intuition and so on, suppose that you are very much more certain of Theory A than of Theory B—much more sure. But suppose that the thing that you are going to observe is a test. According to Theory A, nothing should happen. According to Theory B, it should turn blue. Well, you make the observation, and it turns sort of a greenish. Then you look at Theory A, and you say, "It's very unlikely," and you turn to Theory B, and you say, "Well, it should have turned sort of blue, but it wasn't impossible that it should turn sort of greenish color." So the result of this observation, then, is that Theory A is getting weaker, and Theory B is getting stronger. And if you continue to make more tests, then the odds on Theory B increase. Incidentally, it is not right to simply repeat the same test over and over and over and over, no matter how many times you look and it still looks greenish, you haven't made up your mind yet. But if you find a whole lot of other things that distinguish Theory A from Theory B that are different, then by accumulating a large number of these, the odds on Theory B increase.

Example. I'm in Las Vegas, suppose. And I meet a mind reader, or, let's say, a man who claims not to be a mind reader, but more technically speaking to have the ability of telekinesis, which means that he can influence the way things behave by pure thought. This fellow comes to me, and he says, "I will demonstrate this to you. We will stand at the roulette wheel and I will tell you ahead of time whether it is going to be black or red on every shot."

I believe, say, before I begin, it doesn't make any difference what number you choose for this. I happen to be prejudiced against mind readers from experience in nature, in physics. I don't see, if I believe that man is made out of atoms and if I know all of the—most of the—ways atoms interact with each other, any direct way in which the machinations in the mind can affect the ball. So from other experience and general knowledge, I have a strong prejudice against mind readers. Million to one.

Now we begin. The mind reader says it's going to be black. It's black. The mind reader says it's going to be red. It's red. Do I believe in mind readers? No. It could happen. The mind reader says it's going to be black. It's black. The mind reader says it's going to be red. It's red. Sweat. I'm about to learn something. This continues, let us suppose, for ten times. Now it's possible by chance that that happened ten times, but the odds are a thousand to one against it. Therefore, I now have to conclude that the odds that a mind reader is really doing it are a thousand to one that he's not a mind reader still, but it was a million to one before. But if I get ten more, you see, he'll convince me. Not quite. One must always allow for alternative theories. There is another theory that I should have mentioned before. As we went up to the roulette table, I must have thought in my mind of the possibility that there is collusion between the so-called mind reader and the people at the table. That's possible. Although this fellow doesn't look like he's got any contact with the Flamingo Club, so I suspect that the odds are a hundred to one against that. However, after he has run ten times favorable, since I was so prejudiced against mind reading, I conclude it's collusion. Ten to one. That it's collu-

cist would love to investigate it as a phenomenon of nature. Does it depend upon how far he is from the ball? What about if you put sheets of glass or paper or other materials in between? That's the way all of these things have been worked out, what magnetism is, what electricity is. And what mind reading is would also be analyzable by doing enough experiments.

Anyway, there is an example of how to deal with uncertainty and how to look at something scientifically. To be prejudiced against mind reading a million to one does not mean that you can never be convinced that a man is a mind reader. The only way that you can never be convinced that a man is a mind reader is one of two things: If you are limited to a finite number of experiments, and he won't let you do any more, or if you are infinitely prejudiced at the beginning that it's absolutely impossible.

Now, another example of a test of truth, so to speak, that works in the sciences that would probably work in other fields to some extent is that if something is true, really so, if you continue observations and improve the effectiveness of the observations, the effects stand out more obviously. Not less obviously. That is, if there is something really there, and you can't see good because the glass is foggy, and you polish the glass and look clearer, then it's more obvious that it's there, not less.

I give an example. A professor, I think somewhere in Virginia, has done a lot of experiments for a number of years on the subject of mental telepathy, the same kind of stuff as mind reading. In his early experiments the game was to have a set of cards with various designs on them (you probably know all this, because they sold the cards and people used to play this game), and you would guess whether it's a circle or a triangle and so on while someone else was thinking about it. You would sit and not see the card, and he would see the card and think about the card and you'd guess what it was. And in the beginning of these researches, he found very remarkable effects. He found people who would guess ten to fifteen of the cards correctly, when it should be on the average only five. More even than that. There were some who would come very close to a hundred percent in going through all the cards. Excellent mind readers.

A number of people pointed out a set of criticisms. One thing, for example, is that he didn't count all the cases that didn't work. And he just took the few that did, and then you can't do statistics anymore. And then there were a large number of apparent clues by which signals inadvertently, or advertently, were being transmitted from one to the other.

Various criticisms of the techniques and the statistical methods were made by people. The technique was therefore improved. The result was that, although five cards should be the average, it averaged about six and a half cards over a large number of tests. Never did he get anything like ten or fifteen or twenty-five cards. Therefore, the phenomenon is that the first experiments are wrong. The second experiments proved that the phenomenon observed in the first experiment was nonexistent. The fact that we have six and a half instead of five on the average now brings up a new possibility, that there is such a thing as mental telepathy, but at a much lower level. It's a different idea, because, if the thing was really there before, having improved the methods of experiment, the phenomenon would still be there. It would still be fifteen cards. Why is it down to six and a half? Because the technique improved. Now it still is that the six and a half is a little bit higher than the average of statistics, and various people criticized it more subtly and noticed a couple of other slight effects which might account for the results. It turned out that people would get tired during the tests, according to the professor. The evidence showed that they were getting a little bit lower on the average number of agreements. Well, if you take out the cases that are low, the laws of statistics don't work, and the average is a little higher than the five, and so on. So if the man was tired, the last two or three were thrown away. Things of this nature were improved still further. The results were that mental telepathy still exists, but this time at 5.1 on the average, and therefore all the experiments which indicated 6.5 were false. Now what about the five? . . . Well, we can go on forever, but the point is that there are always errors in experiments that are subtle and unknown. But the reason that I do not believe

after the investigation, then you could change your mind.

Another principle of the same general idea is that the effect we are describing has to have a certain permanence or constancy of some kind, that if a phenomenon is difficult to experiment with, if seen from many sides, it has to have some aspects which are more or less the same.

If we come to the case of flying saucers, for example, we have the difficulty that almost everybody who observes flying saucers sees something different, unless they were previously informed of what they were supposed to see. So the history of flying saucers consists of orange balls of light, blue spheres which bounce on the floor, gray fogs which disappear, gossamer-like streams which evaporate into the air, tin, round flat things out of which objects come with funny shapes that are something like a human being.

If you have any appreciation for the complexities of nature and for the evolution of life on earth, you can understand the tremendous variety of possible forms that life would have. People say life can't exist without air, but it does under water; in fact it started in the sea. You have to be able to move around and have nerves. Plants have no nerves. Just think a few minutes of the variety of life that there is. And then you see that the thing that comes out of the saucer isn't going to be anything like what anybody describes. Very unlikely. It's very unlikely that flying saucers would arrive here, in this particular era, without having caused something of a stir earlier. Why didn't they come earlier? Just when we're getting scientific enough to appreciate the possibility of traveling from one place to another, here come the flying saucers.

There are various arguments of a not complete nature that indicate some doubt that the flying saucers are coming from Venus—in fact, considerable doubt. So much doubt that it is going to take a lot of very accurate experiments, and the lack of consistency and permanency of the characteristics of the observed phenomenon means that it isn't there. Most likely. It's not worth paying much more attention to, unless it begins to sharpen up.

I have argued flying saucers with lots of people. (Incidentally, I must explain that because I am a scientist does not mean that I have not had contact with human beings. Ordinary human beings. I know what they are like. I like to go to Las Vegas and talk to the show girls and the gamblers and so on. I have banged around a lot in my life, so I know about ordinary people.) Anyway, I have to argue about flying saucers on the beach with people, you know. And I was interested in this: they keep arguing that it is possible. And that's true. It is possible. They do not appreciate that the problem is not to demonstrate whether it's possible or not but whether it's going on or not. Whether it's probably occurring or not, not whether it could occur.

That brings me to the fourth kind of attitude toward ideas, and that is that the problem is *not* what is possible. That's not the problem. The problem is what is probable, what is happening. It does no good to demonstrate again and again that you can't disprove that this could be a flying saucer. We have to guess ahead of time whether we have to worry about the Martian invasion. We have to make a judgment about whether it is a flying saucer, whether it's reasonable, whether it's likely. And we do that on the basis of a lot more experience than whether it's just possible, because the number of things that are possible is not fully appreciated by the average individual. And it is also not clear, then, to them how many things that are possible must not be happening. That it's impossible that everything that is possible is happening. And there is too much variety, so most likely anything that you think of that is possible isn't true. In fact that's a general principle in physics theories: no matter what a guy thinks of, it's almost always false. So there have been five or ten theories that have been right in the history of physics, and those are the ones we want. But that doesn't mean that everything's false. We'll find out.

To give an example of a case in which trying to find out what is possible is mistaken for what is probable, I could consider the beatification of Mother Seaton. There was a saintly woman who did very many good works for many people. There is no doubt about that—excuse me, there's very little doubt about that. And it has already been announced that she has demonstrated heroicity of virtues. At that stage in the Catholic system for determining saints, the next question is to consider miracles. So the next problem we have is to decide whether she performed miracles.

There was a girl who had acute leukemia, and the doctors don't know how to cure her. In the duress and troubles of the family in the last minutes, many things are tried—different medicines, all kinds of things. Among other things is the possibility of pinning a ribbon which has touched a bone of Mother Seaton to the sheet of the girl and also arranging that several hundred people pray for her health. And the result is that she—no, not the result—then she gets better from leukemia.

A special tribunal is arranged to investigate this. Very formal, very careful, very scientific. Everything has to be just so. Every question has to be asked very carefully. Everything that is asked is written down in a book very carefully. There are a thousand pages of writing, translated into Italian when it got to the Vatican. Wrapped in special strings, and so on. And the tribunal asks the doctors in the case what this was like. And they all agreed that there was no other case, that this was completely unusual, that at no time before had somebody with this kind of leukemia had the disease stopped for chicken pox just before she got better. Has that got anything to do with it? So there is a definite clinical way to test what it is that might have something to do with it—by making comparisons and so forth. The problem is not to determine that something surprising happens. The problem is to make really good use of that to determine what to do next, because if it does turn out that it has something to do with the prayers of Mother Seaton, then it is worthwhile exhuming the body, which has been done, collecting the bones, touching many ribbons to the bones, so as to get secondary things to tie on other beds.

I now turn to another kind of principle or idea, and that is that there is no sense in calculating the probability or the chance that something happens after it happens. A lot of scientists don't even appreciate this. In fact, the first time I got into an argument over this was when I was a graduate student at Princeton, and there was a guy in the psychology department who was running rat races. I mean, he has a T-shaped thing, and the rats go, and they go to the right, and the left, and so on. And it's a general principle of psychologists that in these tests they arrange so that the odds that the things that happen happen by chance is small, in fact, less than one in twenty. That means that one in twenty of their laws is probably wrong. But the statistical ways of calculating the odds, like coin flipping if the rats were to go randomly right and left, are easy to work out. This man had designed an experiment which would show something which I do not remember, if the rats always went to the right, let's say. I can't remember exactly. He had to do a great number of tests, because, of course, they could go to the right accidentally, so to get it down to one in twenty by odds, he had to do a number of them. And it's hard to do, and he did his number. Then he found that it didn't work. They went to the right, and they went to the left, and so on. And then he noticed, most remarkably, that they alternated, first right, then left, then right, then left. And then he ran to me, and he said, "Calculate the probability for me that they should alternate, so that I can see if it is less than one in twenty." I said, "It probably is less than one in twenty, but it doesn't count." He said, "Why?" I said, "Because it doesn't make any sense to calculate after the event. You see, you found the peculiarity, and so you selected the peculiar case."

For example, I had the most remarkable experience this evening. While coming in here, I saw license plate ANZ 912. Calculate for me, please, the odds that of all the license plates in the state of Washington I should happen to see ANZ 912. Well, it's a ridiculous thing. And, in the same way, what he must do is this: The fact that the rat directions alternate suggests the possibility that rats alternate. If he wants to test this hypothesis, one in twenty, he cannot do it from the same data that gave him the clue. He must do another experiment all over again and then see if they alternate. He did, and it didn't work.

The Meaning of It All

Many people believe things from anecdotes in which there is only one case instead of a large number of cases. There are stories of different kinds of influences. Things that happened to people, and they all remember, and how do you explain that, they say. I can remember things in my life, too. And I give two examples of most remarkable experiences.

The first was when I was in a fraternity at M.I.T. I was upstairs typewriting a theme on something about philosophy. And I was completely engrossed, not thinking of anything but the theme, when all of a sudden in a most mysterious fashion, there swept through my mind the idea: my grandmother has died. Now, of course, I exaggerate slightly, as you should in all such stories. I just sort of half got the idea for a minute. It wasn't something strong, but I exaggerate slightly. That's important. Immediately after that the telephone rang downstairs. I remember this distinctly for the reason you will now hear. The man answered the telephone, and he called, "Hey, Pete!" My name isn't Peter. It was for somebody else. My grandmother was perfectly healthy, and there's nothing to it. Now what we have to do is to accumulate a large number of these in order to fight the few cases when it could happen. It could happen. It might have occurred. It's not impossible, and from then on am I supposed to believe in the miracle that I can tell when my grandmother is dying from something in my head? Another thing about these anecdotes is that all the conditions are not described. And for that reason I describe another, less happy, circumstance.

I met a girl at about thirteen or fourteen whom I loved very much, and we took about thirteen years to get married. It's not my present wife, as you will see. And she got tuberculosis and had it, actually, for several years. And when she got tuberculosis I gave her a clock which had nice big numbers that turned over rather than ones with a dial, and she liked it. The day she got sick I gave it to her, and she kept it by the side of her bed for four, five, six years while she got sicker and sicker. And ultimately she died. She died at 9:22 in the evening. And the clock stopped at 9:22 in the evening and never went again. Fortunately, I noticed some part of the anecdote I have to tell you. After five years the clock gets kind of weak in the knees. Every once in a while I had to fix it, so the wheels were loose. And secondly, the nurse who had to write on the death certificate the time of death, because the light was low in the room, took the clock and turned it up a little bit to see the numbers a little bit better and put it down. If I hadn't noticed that, again I would be in some trouble. So one must be very careful in such anecdotes to remember all the conditions, and even the ones that you don't notice may be the explanation of the mystery.

So, in short, you can't prove anything by one occurrence, or two occurrences, and so on. Everything has to be checked out very carefully. Otherwise you become one of these people who believe all kinds of crazy stuff and doesn't understand the world they're in. Nobody understands the world they're in, but some people are better off at it than others.

The next kind of technique that's involved is statistical sampling. I referred to that idea when I said they tried to arrange things so that they had one in twenty odds. The whole subject of statistical sampling is somewhat mathematical, and I won't go into the details. The general idea is kind of obvious. If you want to know how many people are taller than six feet tall, then you just pick people out at random, and you see that maybe forty of them are more than six feet so you guess that maybe everybody is. Sounds stupid. Well, it is and it isn't. If you pick the hundred out by seeing which ones come through a low door, you're going to get it wrong. If you pick the hundred out by looking at your friends you'll get it wrong because they're all in one place in the country. But if you pick out a way that as far as anybody can figure out has no connection with their height at all, then if you find forty out of a hundred, then, in a hundred million there will be more or less forty million. How much more or how much less can be worked out quite accurately. In fact, it turns out that to be more or less correct to 1 percent, you have to have 10,000 samples. People don't realize how difficult it is to get the accuracy high. For only 1 or 2 percent you need 10,000 tries.

The people who judge the value of advertising in television use this method. No, they think they use this method. It's a very difficult thing to do, and the most difficult part of it is the choice of the samples. How they can arrange to have an average guy put into his house this gadget by which they remember which TV programs he's looking at, or what kind of a guy an average guy is who will agree to be paid to write in a log, and how accurately he writes in the log what he's listening to every fifteen minutes when a bell goes off, we don't know. We have no right, therefore, to judge from the thousand, or 10,000, and that's all it is, people who do this, who study what the average person is looking at, because there's no question at all that the sample is off. This business of statistics is well known, and the problem of getting a good sample is a very serious one, and everybody knows about it, and it's a scientifically OK business. Except if you don't do it. The conclusion from all the researchers is that all people in the world are as dopey as can be, and the only way to tell them anything is to perpetually insult their intelligence. This conclusion may be correct. On the other hand, it may be false. And we are making a terrible mistake if it is false. It is, therefore, a matter of considerable responsibility to get straightened out on how to test whether or not people pay attention to different kinds of advertising.

As I say, I know a lot of people. Ordinary people. And I think their intelligence is being insulted. I mean there's all kinds of things. You turn on the radio; if you have any soul, you go crazy. People have a way—I haven't learned it yet—of not listening to it. I don't know how to do it. So in order to prepare this talk I turned on the radio for three minutes when I was at home, and I heard two things.

First, I turned it on and I heard Indian music—Indians from New Mexico, Navajos. I recognized it. I had heard them in Gallup, and I was delighted. I won't give an imitation of the war chant, although I would like to. I'm tempted. It's very interesting, and it's deep in their religion, and it's something that they respect. So I would report honestly that I was pleased to see that on the radio there was something interesting. That was cultural. So we have to be honest. If we're going to report, you listen for three minutes, that's what you hear. So I kept listening. I have to report that I cheated a little bit. I kept listening because I liked it; it was good. It stopped. And a man said, "We are on the warpath against automobile accidents." And then he went on and said how you have to be careful in automobile accidents. That's not an insult to intelligence; it's an insult to the Navajo Indians, and to their religion and their ideas. And so I listened until I heard that there is a drink of some kind, I think it's called Pepsi-Cola, for people who think young. So I said, all right, that's enough. I'll think about that a while. First of all, the whole idea is crazy. What is a person who thinks young? I suppose it is a person who likes to do things that young people like to do. Alright, let them think that. Then this is a drink for such people. I suppose that the people in the research department of the drink company decided how much lime to put in as follows: "Well, we used to have a drink that was just an ordinary drink, but we have to rearrange it, not for ordinary people but for special people who think young. More sugar." The whole idea that a drink is especially for people who think young is an absolute absurdity.

So as a result of this, we get perpetually insulted, our intelligence always insulted. I have an idea of how to beat it. People have all kinds of plans, you know, and the F.T.C. is trying to straighten it out. I've got an easy plan. Suppose that you purchased the use for thirty days of twenty-six billboards in Greater Seattle, eighteen of them lighted. And you put onto the billboards a sign which says, "Has your intelligence been insulted? Don't buy the product." And then you buy a few spots on the television or the radio. In the middle of some program a man comes up and says, "Pardon me, I'm sorry to interrupt you, but if you find that any of the advertising that you hear insults your intelligence or in any way disturbs you, we would advise you not to buy the product," and things will be straightened out as quickly as it can be. Thank you.

Now if anybody has any money that they want to throw around, I'd advise that as an experiment to find out about the intelligence of the average television looker. It's an interesting question. It's a quick shortcut to find out about their intelligence. But maybe it's a little bit expensive.

You say, "It's not very important. The advertisers have to sell their wares," and so on and so on. On the other hand, the whole idea that the average person is unintelligent is a very dangerous idea. Even if it's true, it shouldn't be dealt with the way it's dealt with.

Newspaper reporters and commentators—there is a large number of them who assume that the public is stupider than they are, that the public cannot understand things that they [the reporters and the commentators] cannot understand. Now that is ridiculous. I'm not trying to say they're dumber than the average man, but they're dumber in some way than somebody else. If I ever have to explain something scientific to a reporter, and he says what is the idea? Well, I explain it in words of one syllable, as I would explain it to my neighbor. He doesn't understand it, which is possible, because he's brought up differently—he doesn't fix washing machines, he doesn't know what a motor is, or something. In other words, he has no technical experience. There are lots of engineers in the world. There are lots of mechanically minded people. There are lots of people who are smarter than the reporter, say, in science, for example. It is, therefore, his duty to report the thing, whether he understands it or not, accurately and in the way it's been given. The same goes in economics and other situations. The reporters appreciate the fact that they don't understand the complicated business about international trade, but they report, more or less, what somebody says, pretty closely. But when it comes to science, for some reason or another, they will pat me on the head and explain to dopey me that the dopey people aren't going to understand it because he, dope, can't understand it. But I *know* that some people can understand it. Not everybody who reads the newspaper has to understand every article in the newspaper. Some people aren't interested in science. Some are. At least they could find out what it's all about instead of discovering that an atomic bullet was used that came out of a machine that weighed seven tons. I can't read the articles in the paper. I don't know what they mean. I don't know what kind of a machine it was just because it weighed seven tons. And there are now sixty-two kinds of particles, and I would like to know what atomic bullet he is referring to.

This whole business of statistical sampling and the determining of the properties of people by this manner is a very serious business altogether. It's coming into its own, but it's used very often, and we have to be very, very careful with it. It's used for choice of personnel—by giving examinations to people—marriage counseling, and things of this kind. It's used to determine whether people get into college, in a way that I am not in favor of, but I will leave my arguments on this. I will address them to the people who decide who gets into Caltech. And after

I have had my arguments, I will come back and tell you something about it. But this has one serious feature, among others, aside from the difficulties of sampling. There is a tendency, then, to use only what can be measured as a criterion. That is, the spirit of the man, the way he feels toward things, may be difficult to measure. There is some tendency to have interviews and to try to correct this. So much the better. But it's easier to have more examinations and not have to waste the time with the interviews, and the result is that only those things which can be measured, actually which they think they can measure, are what count, and a lot of good things are left out, a lot of good guys are missed. So it's a dangerous business and has to be very carefully checked. The things like marriage questions, "How are you getting along with your husband," and so on, that appear in magazines are all nonsense. They go something like this: "This has been tested on a thousand couples." And then you can tell how they answered and how you answered and tell if you are happily married. What you do is the following. You make up a bunch of questions, like "Do you give him breakfast in bed?" and so on and so on. And then you give this questionnaire to a thousand people. And you have an independent way of telling whether they are happily married, like asking them, or something. But never mind. It doesn't make any difference what it is, even if the test is perfect. That's not the part where the trouble is. Then you do the following. You see about all the ones who are happy—how did they answer about the breakfast in bed, how did they answer about this, how did they answer about that? You see it's exactly the same as my rat race, with right and left. They have decided on the odds of the thing in terms of the one sample. What they ought to do to be honest is to take the same test that has now been designed, in which they know how to make the score. They've decided this gets five points, that gets ten points, in such a way that the thousand that they tried it on get marvelous scores if they are happy and lousy scores if they're not. But now is the test of the test. They cannot use the sample which determined the scoring for them. That's going backwards. They must take the test to another thousand people, independently, and run it out to see whether the happy ones are the ones that score high, or not. They do not do that, because it's too much trouble, A, and the few times that they tried it, B, it showed that the test was no good.

Now, looking at the troubles that we have with all the unscientific and peculiar things in the world, there are a number of them which cannot be associated with difficulties in how to think, I think, but are just due to some lack of information. In particular, there are believers in astrology, of which, no doubt, there are a number here. Astrologists say that there are days when it's better to go to the dentist than other days. There are days when it's better to fly in an airplane, for you, if you are born on such a day and such and such an hour. And it's all calculated by very careful rules in terms of the position of the stars. If it were true it would be very interesting. Insurance people would be very interested to change the insurance rates on people if they follow the astrological rules, because they have a better chance when they are in the airplane. Tests to determine whether people who go on the day that they are not supposed to go are worse off or not have never been made by the astrologers. The question of whether it's a good day for business or a bad day for business has never been established. Now what of it?

Maybe it's still true, yes. On the other hand, there's an awful lot of information that indicates that it isn't true. Because we have a lot of knowledge about how things work, what people are, what the world is, what those stars are, what the planets are that you are looking at, what makes them go around more or less, where they're going to be in the next 2000 years is completely known. They don't have to look up to find out where it is. And furthermore, if you look very carefully at the different astrologers they don't agree with each other, so what are you going to do? Disbelieve it. There's no evidence at all for it. It's pure nonsense. The only way you can believe it is to have a general lack of information about the stars and the world and what the rest of the things look like. If such a phenomenon existed it would be most remarkable, in the face of all the other phenomena that exist, and unless someone can demonstrate it to you with a real experiment, with a real test, took people who believe and people who didn't believe and made a test, and so on, then there's no point in listening to them. Tests of this kind, incidentally, have been made in the early days of science. It's rather interesting. I found out that in the early days, like in the time when they were discovering oxygen and so on, people made such experimental attempts to find out, for example, whether missionaries—it sounds silly; it only sounds silly because you're afraid to test it—whether good people like missionaries who pray and so on were less likely to be in a shipwreck than others. And so when missionaries were going to far countries, they checked in the shipwrecks whether the missionaries were less likely to drown than other people. And it turned out that there was no difference. So lots of people don't believe that it makes any difference.

There are, if you turn on the radio—I don't know how it is up here; it must be the same—in California you hear all kinds of faith healers. I've seen them on television. It's another one of those things that it exhausts me to try to explain why it's rather a ridiculous proposition. There is, in fact, an entire religion that's respectable, so called, that's called Christian Science, that's based on the idea of faith healing. If it were true, it could be established, not by the anecdotes of a few people but by the careful checks, by the technically good clinical methods which are used on any other way of curing diseases. If you believe in faith healing, you have a tendency to avoid other ways of getting healed. It takes you a little longer to get to the doctor, possibly. Some people believe it strongly enough that it takes them longer to get to the doctor. It's possible that the faith healing isn't so good. It's possible—we are not sure—that it isn't. And it's therefore possible that there is some danger in believing in faith healing, that it's not a triviality, not like astrology wherein it doesn't make a lot of difference. It's just inconvenient for the people who believe in it that they have to do things on certain days. It may be, and I would like to know—it should be investigated—everybody has a right to know—whether more people have been hurt or helped by believing in Christ's ability to heal; whether there is more healing or harming by such a thing. It's possible either way. It should be investigated. It shouldn't be left lying for people to believe in without an investigation.

Not only are there faith healers on the radio, there are also radio religion people who use the Bible to predict all kinds of phenomena that are going to happen. I listened intrigued to a man who in a dream visited God and received all kinds of special information for his congregation, etc. Well, this unscientific age . . . But I don't know what to do with that one. I don't know what rule of reasoning to use to show right away that it's nutty. I think it just belongs to a general lack of understanding of how complicated the world is and how elaborate and how unlikely it would be that such a thing would work.

But I can't disprove, of course, without investigating more carefully. Maybe one way would be always to ask them how do they know it's true and to remember maybe that they are wrong. Just remember that much anyway, because you may keep yourself from sending in too much money.

There are also, of course, in the world a number of phenomena that you cannot beat that are just the result of a general stupidity. And we all do stupid things, and we know some people do more than others, but there is no use in trying to check who does the most. There is some attempt to protect this by government regulation, to protect this stupidity, but it doesn't work a hundred percent.

For example, I went on a visit to one of the desert sites to buy land. You know they sell land, these pro-moters—there's a new city going to be built. It's exciting. It's marvelous. You must go. Just imagine yourself in a desert with nothing but some flags poked here in the ground with numbers on them and street signs with names. And so you drive in the car across the desert to find the fourth street and so on to get to the lot 369, which is the one for you, you're thinking. And you stand there kicking sand in this thing discussing with the sales-man why it's advantageous to have a corner lot and how the driveway will be good because it will be easier to get into from that side. Worse, believe it or not, you find yourself discussing the beach club, which is going to be on that sea, what the rules of membership are and how many friends you're allowed to bring. I swear, I got into that condition.

So when the time comes to buy the land, it turns out that the state has made an attempt to help you. So they have a description of this particular thing that you have read, and the man who sells you the land says it's the law, we have to give you this to read. They give it to you to read, and it says that this is very much like many other real estate deals in the state of California and so on and so on and so on. And among other things, I read that although they say that they want to have fifty thousand people at this site, there is not water enough for a number which I better not say or I'll get accused of libel, but it was very much less—I can't remember it exactly—it was in the neighborhood of five thousand people, somewhere like that. So, of course they had noticed that this was in there before, and they told us that they had just found water at another site, far away, that they were going to pump down. And when I asked about it, they explained to me very carefully that they had just discovered this and that they hadn't had time to get it into the brochure from the state. Hmmmm.

I'll give another example of the same thing. I was in Atlantic City, and I went into one of these—well, it was sort of a store. There were a lot of seats, and people were sitting there listening to a man speaking. And he was very interesting. He knew all about food, and he was talking about nutrition, different things. I remember several of the important statements which he made, such as "even worms won't eat white flour." That kind of stuff. It was good. It was interesting. It was true—maybe it wasn't true about the worms, but it was good stuff about proteins and so on. And then he went on and described the Federal Pure Food and Drug Act, and he explained how it protects you. He explained that on every product that claims to be a good health food that's supposed to help you with minerals and this and that, there must be a label on the bottle which tells exactly what's in it, what it does, and all claims must be explicit, so that if it's wrong, so on and so on. He gives them everything. I said, "How is he going to make any money? Out come the bottles. It comes out, finally, that he sells this special health food, of course, in a brownish bottle. And it just so happens that he has just come in, and he's been in a hurry, and he hasn't had time to put the labels on. And here are the labels that belong on the bottles, and here are the bottles, and he's in a hurry to sell them, and he gives you the bottle, and you stick it on yourself. That man had courage. He first explained what to do, what to worry about, and then he went ahead and did it.

I found another lecture which was somewhat analogous to that one. And that was the second Danz lecture given by myself. I started out by pointing out that things were completely unscientific, that things were uncertain, particularly in political matters, and that there were the two nations, Russia and the United States, at odds with each other. And then by some mystic hocus-pocus it came out that we were the good guys and they were the bad guys. Yet, at the beginning, there was no way to decide which was the better of the two. In fact, that was the main point of the lecture. So by some sort of magic I produced some kind of relative certainty out of uncertainty. I told you about the bottle with the labels, and then I came out on the other end with a label on my bottle. How did I do it? You have to think about it a little bit. One thing, of course, that we can be certain of, once we're uncertain, and that is that we are uncertain. Somebody says "No, maybe I'm sure." Actually, though, the gimmick in that particular lecture, the weak point in the whole thing, the thing that requires further development and study is this one: I made an impassioned plea for the idea that it's good to have an open channel, that there's value in uncertainty, that it's more important to permit us to discover new things, rather than to choose a solution that we now make up—that to choose a solution, no matter how we choose it now is to choose a much worse thing than what we would get if we waited and worked things out. And that's where I made the choice, and I am not sure of that choice. Okay. I have now destroyed authority.

Associated with these problems of lack of information and so forth, but particularly lack of information, there are a number of phenomena that are more serious, I believe, than astrology.

I, in preparation for this lecture, investigated something that was in my town, in the shopping center. There was a store with a flag in front. And it's the Americanism Center, Altadena Americanism Center. And so I went into the Americanism Center to find out what it is, and it's a volunteer organization. And on the front outside, there is a Constitution and the Bill of Rights and so on, and a letter which explains their purpose, which is to maintain rights and so on, all in accordance with the Constitution and the Bill of Rights and so on. That's the general idea. What they do in there is simply educative. They have books that people could buy on the various subjects that help to teach the ideas of citizenship and so on, and they have, among other books, also Congressional records, pamphlets on Congressional investigations and so on, so that people who are studying these problems can read them. They have study groups which meet at night, and so on. So, being interested in rights for people, I asked, since I said I didn't know very much about it, I would like a book on the problem of the freedom of the Negroes to vote in the South. There was nothing. Yes, there was. There was one thing which turned up later, two things which I saw out of the corner of my eye. One was what went on in Mississippi according to the Oxford city fathers, and the other was a little pamphlet called "The National Association for the Advancement of Colored People and Communism."

So I discussed it at some greater length to discover

what was going on and talked to the lady for a while, and she explained among other things (we talked about many things—we did this on a friendly basis, you will be surprised to hear) that she was not a member of the Birch Society but there was something that you could say for the Birch Society, she saw some movie about it and so on, and there was something that she could say for it. You're not a fence sitter when you're in the Birch Society. At least you know what you're for, because you don't have to join it if you don't want to, and this is what Mr. Welch said, and this is the way the Birch Society is, and if you believe in this then you join, and if you don't believe in this then you shouldn't join. It sounds just like the Communist Party. It's all very well if they have no power. But if they have power, it's a completely different situation. I tried to explain to her that this is not the kind of freedom that was being talked about, that in any organization there ought to be the possibility of discussion. That fence sitting is an art, and it's difficult, and it's important to do, rather than to go headlong in one direction or the other. It's just better to have action, isn't it, than to sit on the fence? Not if you're not sure which way to go, it isn't.

So I bought a couple of things there, just at random that they had. One of the things was called "The Dan Smoot Report"—it's a good name—and it talked about the Constitution, and a general idea I'll outline: that the Constitution was right the way it was written in the first place. And all the modifications that have come in are just the mistakes. Fundamentalists, only not in the Bible but in the Constitution. And then it goes on to give the ratings of Congressmen in votes, how they voted. And it said, very specifically and after explaining about their ideas, "The following give the ratings of the congressmen and senators with regard to whether they vote for or against the Constitution." Mind you that these ratings are not just an opinion, but they are based on fact. They are a matter of voting record. Fact. There's no opinion at all. It's just the voting record, and, of course, each item is either for or against the Constitution. Naturally. Medicare is against the Constitution, and so on. I tried to explain that they violate their own principles. According to the Constitution there are supposed to be votes. It isn't supposed to be automatically determinable ahead of time on each one of the items what's right and what's wrong. Otherwise there wouldn't be the bother to invent the Senate to have the votes. As long as you have the votes at all, then the purpose of the votes is to try to make up your mind which is the way to go. And it isn't possible for somebody to determine by fact ahead of time what is the situation. It violates its own principle.

It starts out all right, with the good, and love, and Christ, and so on, and it builds itself up until it's afraid of an enemy. And then it forgets its original idea. It turns itself inside out and becomes absolutely contrary to the beginning. I believe that the people who start some of these things, especially the volunteer ladies of

Altadena, have a good heart and understand a little bit that it's good, the Constitution, and so on, but they are led astray in the system of the thing. How, I can't exactly get at, and what to do to keep from doing this, I don't exactly know.

I went still further into the thing and found out what the study group was about, and if you don't mind I'll tell you what that was about. They gave me some papers. There were a lot of chairs, you see, in the room, and they explained to me, yes, that evening they had a study group, and they gave me a thing which described what they were going to study. And I made some notes from it. It had to do with the S.P.X.R.A. In 1943 the S.P.X. research associates—which turns out to be the . . . well, I'll tell you what it turns out to be—came into being through the professional interest of intelligence officers then on active duty in the armed forces of the United States concerning the Soviet revival of a long dormant tenth principle of warfare. Paralysis. See the evil. Dormant. Mysterious. Frightening. The mystic people of the military orders have had principles of warfare since the Roman legions. Number one. Number two. Number three. This is number ten. We don't have to know what number seven is. The whole idea that there are long dormant principles of warfare, much less that there is a tenth principle of warfare, is an absurdity. And then what is this principle of paralysis? How are they going to use the idea? The boogie man is now generated. How do you use the boogie man? You use the boogie man as follows: This educational program concerns itself with all the areas where Soviet pressure can be used to paralyze the American will to resist. Agriculture, arts, and cultural exchange. Science, education, information media, finance, economics, government, labor, law, medicine, and our armed forces, and religion, that most sensitive of areas. In other words, we now have an open machine for pointing out that everybody who says something that you don't agree with has been paralyzed by the mystic force of the tenth principle of warfare.

This is a phenomenon analogous to paranoia. It is impossible to disprove the tenth principle. It's only possible if you have a certain balance, a certain understanding of the world to appreciate that it's out of balance, to think that the Supreme Court—which turns out to be an "instrument of global conquest"—has been paralyzed. Everything is paralyzed. You see how fearful it becomes, the terrible power which is demonstrated again and again by one example after the other of this fearful force which is made up.

This describes what a paranoia is like. A woman gets nervous. She begins to suspect that her husband is trying to make trouble for her. She doesn't like to let him into the house. He tries to get into the house, proves that he's trying to make trouble for her. He gets a friend to try to talk to her. She knows that it's a friend, and she knows in her mind, which is going to one side, that this is only further evidence of the terrible fright and the fear that she's building up in her mind. Her neighbors come over to console her for a while. It works fairly well, for a while. They go back to their houses. The friend of the husband goes to visit them. They are spoiled now, and they are going to tell her husband all the terrible things she said. Oh dear, what did she say? And he's going to be able to use them against her. She calls up the police department. She says, "I'm afraid." She's locked in her house now. She says, "I'm afraid." Somebody's trying to get into the house. They come, they try to talk to her, they realize that there is nobody trying to get into the house. They have to go away. She remembers that her husband was important in the city. She remembers that he had a friend in the police department. The police department is only part of the scheme. It only proves it once again. She looks through the window of the house, and she sees across the way someone stopping at a neighbor's house. What are they talking about? In the backyard, she sees something coming up over a bush. They're watching her with a telescope! It turns out later to be some children playing in the back with a stick. A continuous and perpetual buildup, until the entire population is involved. The lawyer that she called, she remembers, was the lawyer once for a friend of her husband's. The doctor who has been trying to get her to the hospital is now obviously on the side of the husband.

The only way out is to have some balance, to think

that it's impossible that the whole city is against her, that everybody is going to pay attention to this husband of mine who's such a dope, that everybody's going to do all these things, that there's a complete accumulation. All the neighbors, everybody's against her. It's out of proportion. It's only out of proportion. How can you explain to somebody who hasn't got a sense of proportion?

And so it is with these people. They don't have a sense of proportion. And so they will believe in such a possibility as the Soviet tenth principle of warfare. The only way that I can think to beat the game is to point the following out. They're right. And like my friend with the bottle with the label, the Soviets are very, very ingenious and clever indeed. They even tell us what they're doing to us. You see, these people, these research associates are really in the hire of the Soviets who are using this method of paralysis. And what they want us to do is to lose faith in the Supreme Court, to lose faith in the Agriculture Department, to lose faith in the scientists and all the people who help us in all kinds of ways and so on and so on, and lose faith in all sorts of ways, and it's a way that they have entered into this movement of freedom that everybody wanted, this thing with all the flags and the Constitution, and they've gotten in on it, and they're getting in there, and they're going to paralyze it. Proof. In their own words. S.P.X.R.A. has qualified, under oath, in the United States court as the leading, American authority on the tenth principle. Where did

don't know myself whether I am for nuclear testing or against nuclear testing. There are reasons on both sides. It makes radioactivity, and it's dangerous, and it's also very bad to have a war. But whether it's going to be more likely to have a war or less likely to have a war because you test, I don't know. Whether preparation will stop the war, or lack of preparation, I don't know. So I'm not trying to say I'm on either side. That's why I can be abjectly honest on this one.

The big question comes, of course, whether there's a danger from radioactivity. In my opinion the greatest danger and the greatest question on nuclear testing is the question of its future effects. The deaths and the radioactivity which would be caused by the war would be so many times more than the nuclear testing that the effects that it would have in the future are far more important than the infinitesimal amount of radioactivity produced now. How infinitesimal is the amount, however? Radioactivity is bad. Nobody knows a good effect of general radioactivity. So if you increase the general amount of radioactivity in the air, you are producing something not good. Therefore nuclear testing in this respect produces something not good. If you are a scientist, then, you have the right and should point out this fact.

On the other hand, the thing is quantitative. The question is how much is not good? You can play games and show that you will kill 10 million people in the next 2000 years with it. If I were to walk in front of a car, hoping that I will have some more children in the future, I also will kill 10,000 people in the next 10,000 years, if you figure it out, from a certain way of calculating. The question is how big is the effect? And the last time . . . (I wish I had—I should, of course, have checked these figures, but let me put it differently.) The next time you hear a talk, ask the questions which I point out to you, because I asked some questions the last time I heard a talk, and I can remember the answers, but I haven't checked them very recently, so I don't have any figures, but I at least asked the question. How much is the increase in radioactivity compared to the general variations in the amount of radioactivity from place to place? The amounts of background radioactivity in a wooden building and a brick building are quite different, because the wood is less radioactive than the bricks.

It turns out that at the time that I asked this question, the difference in the effects was less than the difference between being in a brick and a wooden building. And the difference between being at sea level and being at 5000 feet altitude was a hundred times, at least, bigger than the extra radioactivity produced by the atomic bomb testing.

Now, I say that if a man is absolutely honest and wants to protect the populace from the effects of radioactivity, which is what our scientific friends often say they are trying to do, then he should work on the biggest number, not on the smallest number, and he should try to point out that the radioactivity which is produced by living in the city of Denver is so much more serious, is a hundred times bigger than the background from the bomb, that all the people of Denver ought to move to lower altitudes. The situation really is—don't get frightened if you live in Denver—it's small. It doesn't make much difference. It's only a tiny effect. But the effect from the bombs is less than the difference between being at low level and high level, I believe. I'm not absolutely sure. I ask you to ask that question to get some idea whether you should be very careful about not walking into a brick building, as careful as you are to try to stop nuclear testing for the sole reason of radioactivity. There are many good reasons that you may feel politically strong about, one way or the other. But that's another question.

We are, in the scientific things, getting into situations in which we are related to the government, and we have all kinds of lack of honesty. Particularly, lack of honesty is in the reporting and description of the adventures of going to different planets and in the various space adventures. To take an example, we can take the Mariner II voyage to Venus. A tremendously exciting thing, a marvelous thing, that man has been able to send a thing 40 million miles, a piece of the earth at last to another place. And to get so close to it as to get a view that corresponds to being 20,000 miles away. It's hard for me to explain how exciting that is, and how interesting. And I've used up more time than I ought.

The story of what happened during the trip was equally interesting and exciting. The apparent breakdown. The fact that they had to turn all the instruments off for a while because they were losing power in the batteries and the whole thing would stop. And then they were able to turn it on again. The fact of how it was heating up. How one thing after the other didn't work and then began to work. All the accidents and the excitement of a new adventure. Just like sending Columbus, or Magellan, around the world. There were mutinies, and there were troubles and there were shipwrecks, and there was the whole works. And it's an exciting story. When it, for example, heated up, it was said in the paper, "It's heating up, and we're learning from that." What could we be learning? If you know something, you realize you can't learn anything. You put satellites up near the earth, and you know how much radiation you get from the sun . . . we know that. And how much do they get when they get near Venus? It's a definitely accurate law, well known, inverse square. The closer you get, the brighter the light. Easy. So it's easy to figure out how much white and black to paint the thing so that the temperature adjusts itself.

The only thing we learned was that the fact that it got hot was not due to anything else than the fact that the thing was made in a very great hurry at the last minute and some changes were made in the inside apparatus, so that there was more power developed in the inside and it got hotter than it was designed for. What we learned, therefore, was not scientific. But we learned to be a little bit careful about going in such a hurry on these things and keep changing our minds at the last minute. By some miracle the thing almost worked when it was there. It was meant to look at Venus by making a series of passes across the planet, looking like a television screen, twenty-one passes across the planet. It made three. Good. It was a miracle. It was a great achievement. Columbus said he was going for gold and spices. He got no gold and very little spices. But it was a very important and very exciting moment. Mariner was supposed to go for big and important scientific information. It got none. I tell you it got none. Well, I'll correct it in a minute. It got practically none. But it was a terrific and exciting experience. And in the future more will come from it. What it did find out, from looking at Venus, they say in the paper, was that the temperature was 800 degrees or something, under the surface of the clouds. That was already known. And it's being confirmed today, even now, by using the telescope at Palomar and making measurements on Venus from the earth. How clever. The same information could be gotten from looking from the Earth: I have a friend who has information on this, and he has a beautiful map of Venus in his room, with contour lines and hot and cold and different temperatures in different parts. In detail. From the earth. Not just three swatches with some spots of up and reasonable and simple explanation. It's not necessary that we have so many failures, as far as I can tell. There's something the matter in the organization, in the administration, in the engineering, or in the making of these instruments. It's important to know that. It's not worthwhile knowing that we're always learning something.

Incidentally, people ask me, why go to the moon? Because it's a great adventure in science. Incidentally, it also develops technology. You have to make all these instruments to go to the moon—rockets, and so on—and it's very important to develop technology. Also it makes scientists happy, and if scientists are happy maybe they'll work on something else good for warfare. Another possibility is a direct military use of space. I don't know how, nobody knows how, but there may turn out to be a use. Anyway, it's possible that if we keep on developing the military aspects of long-range flying to the moon that we'll prevent the Russians from making some military use that we can't figure out yet. Also there are indirect military advantages. That is, if you build bigger rockets, then you can use them more directly by going directly from here to some other part of the earth instead of having to go to the moon. Another good reason is a propaganda reason. We've lost some face in front of the world by letting the other guys get ahead in technology. It's good to be able to try to get that face back. None of these reasons alone is worthwhile and can explain our going to the moon. I believe, however, that if you put them all

you consider all the structures and inventions and complicated things, the ids and the egos, the tensions and the forces, and the pushes and the pulls, I tell you they can't all be there. It's too much for one brain or a few brains to have cooked up in such a short time. However, I remind you that if you're in the tribe, there's nobody else to go to.

And now I can have some more fun, and this is especially for the students of this university. I thought, among other people, of the Arabian scholars of science during the Middle Ages. They did a little bit of science themselves, yes, but they wrote commentaries on the great men that came before them. They wrote commentaries on commentaries. They described what each other wrote about each other. They just kept writing these commentaries. Writing commentaries is some kind of a disease of the intellect. Tradition is very important. And freedom of new ideas, new possibilities, are disregarded on the grounds that the way it was is better than anything I can do. I have no right to change this or to invent anything or to think of anything. Well, those are your English professors. They are steeped in tradition, and they write commentaries. Of course, they also teach us, some of us, English. That's where the analogy breaks down.

Now if we continue in the analogy here, we see that if they had a more enlightened view of the world there would be a lot of interesting problems. Maybe, how many parts of speech are there? Shall we invent another part of speech? Ooohhhhh!

Well, then how about the vocabulary? Have we got too many words? No, no. We need them to express ideas. Have we got too few words? No. By some accident, of course, through the history of time, we happened to have developed the perfect combination of words.

Now let me get to a lower level still in this question. And that is, all the time you hear the question, "why can't Johnny read?" And the answer is, because of the spelling. The Phoenicians, 2000, more, 3000, 4000 years ago, somewhere around there, were able to figure out from their language a scheme of describing the sounds with symbols. It was very simple. Each sound had a corresponding symbol, and each symbol, a corresponding sound. So that when you could see what the symbols' sounds were, you could see what the words were supposed to sound like. It's a marvelous invention. And in the period of time things have happened, and things have gotten out of whack in the English language. Why can't we change the spelling? Who should do it if not the professors of English? If the professors of English will complain to me that the students who come to the universities, after all those years of study, still cannot spell "friend," I say to them that something's the matter with the way you spell friend.

And also, it can be argued, perhaps, if they wish, that it's a question of style and beauty in the language, and that to make new words and new parts of speech might destroy that. But they cannot argue that respelling the words would have anything to do with the style. There's no form of art form or literary form, with the sole exception of crossword puzzles, in which the spelling makes a bit of difference to the style. And even crossword puzzles can be made with a different spelling. And if it's not the English professors that do it, and if we give them two years and nothing happens—and please don't invent three ways of doing it, just one way, that everybody is used to—if we wait two or three years and nothing happens, then we'll ask the philologists and the linguists and so on because they know how to do it. Did you know that they can write any language with an alphabet so that you can read how it sounds in another language when you hear it? That's really something. So they ought to be able to do it in English alone.

One thing else I would leave to them. This does show, of course, that there are great dangers in arguing from analogy. And these dangers should be pointed out. I don't have time to do that, and so I leave to your English professors the problem of pointing out the errors of reasoning by analogy.

Now there are a number of things, positive things, in which a scientific type of reasoning works, and in which considerable progress has been made, and I've been picking out a number of the negative things. I want you to know I appreciate positive things. (I also appreciate that I'm talking too long, so I will mention them only. But it's out of proportion. I wanted to spend more time.) There are a number of things in which rational people work very hard using methods which are quite sensible. And nobody's bothered with them, yet.

For instance, people have arranged traffic systems and arranged the way the traffic will work in other cities. Criminal detection is at a pretty high level of knowing how to get evidence, how to judge evidence, how to control your emotions on the evidence, and so on.

We shouldn't only think of the technological inventions when we consider the progress of man. There are an enormous number of most important nontechnological inventions which mustn't be disregarded. Economic inventions in checks, for example, and banks, things of this nature. International financial arrangements, and so on, are marvelous inventions. And they are absolutely essential and represent a great advance. Systems of accounting, for example. Business accounting is a scientific process—I mean, is not a scientific, maybe, but a rational process. A system of law has been gradually developed. There is a system of laws and juries and judges. And although there are, of course, many faults and flaws, and we must continue to work on them, I have great admiration for that. And also the development of government organizations which have been going on through the years. There are a large number of problems which have been solved in certain countries in ways that we sometimes can understand and sometimes we cannot. I remind you of one, because it bothers me. And that has to do with the fact that the government really has the problem of the control of the forces. And most of the time there has been trouble because the strongest forces try to get control of the government. It is marvelous, is it not, that someone with no force can control someone with force. And so the difficulties in the Roman empire, with the Praetorian guards, seemed insoluble, because they had more force than the Senate. Yet in our country we have a sort of discipline of the military, so that they never try to control the Senate directly. People laugh at the brass. They tease them all the time. No matter how many things we've stuffed down their throats, we civilians have still been able to control the military! I think that the military's discipline in knowing what its place is in the government of the United States is one of our great heritages and one of the very valuable things, and I don't think that we should keep pushing on them so hard until they get impatient and break out from their self-imposed discipline. Don't misunderstand me. The military has a large number of faults, like anything else. And the way they handled Mr. Anderson, I believe his name was, the fellow who was supposed to have murdered somebody and so on, is an example of what would happen if they did take over.

Now, if I look to the future, I should talk about the future development of mechanics, the possibilities that will arise because we have almost free energy when we get to controlled fusion. And in the near future the developments in biology will make problems like no one has ever seen before. The very rapid developments of biology are going to cause all kinds of very exciting problems. I haven't time to describe them, so I just refer you to Aldous Huxley's book *Brave New World*, which gives some indication of the type of problem that future biology will involve itself in.

One thing about the future I look to with favor. I think there are a lot of things working in the right direction. In the first place, the fact that there are so many nations and they hear each other, on account of the communications, even if they try to close their ears. And so there are all kinds of opinions running around, and the net result is that it's hard to keep ideas out. And some of the troubles that the Russians are having in holding down people like Mr. Nakhrosov are a kind of trouble that I hope will continue to develop.

One other point that I would like to take a moment or two to make a little bit more in detail is this one: The problem of moral values and ethical judgments is one into which science cannot enter, as I have already indicated, and which I don't know of any particular way to word. However, I see one possibility. There may be others, but I see one possibility. You see we need some kind of a mechanism, something like the trick we have to make an observation and believe it, a scheme for choosing moral values. Now in the days of Galileo there were great arguments about what makes a body fall, all kinds of arguments about the medium and the pushes and the pulls and so on. And what Galileo did was disregard all the arguments and decide if it fell and how fast it fell, and just describe that. On that everybody could agree. And keep on studying in that direction, on what everyone can agree, and never mind the machinery and the theory underneath, as long as possible. And then gradually, with the accumulation of experience, you find other theories underneath that are more satisfactory, perhaps. There were in the early days of science terrible arguments about, for instance, light. Newton did some experiments which showed that a light beam separated and spread with a prism would never get separated again. Why did he have to argue with Hooke? He had to argue with Hooke because of the theories of the day about what light was like and so on. He wasn't arguing whether the phenomenon was right. Hooke took a prism and saw that it was true.

So the question is whether it is possible to do something analogous (and work by analogy) with moral problems. I believe that it is not at all impossible that there be agreements on consequences, that we agree on the net result, but maybe not on the reason we do what we ought to do. That the argument that existed in the early days of the Christians as to, for instance, whether Jesus was of a substance *like* the Father or of the *same* substance *as* the

Father, which when translated into the Greek became the argument between the Homoiousions and the Homoousians. Laugh, but people were hurt by that. Reputations were destroyed, people were killed, arguing whether it's the same or similar. And today we should learn that lesson and not have an argument as to the reason *why* we agree if we agree.

I therefore consider the Encyclical of Pope John XXIII, which I have read, to be one of the most remarkable occurrences of our time and a great step to the future. I can find no better expression of my beliefs of morality, of the duties and responsibilities of mankind, people to other people, than is in that encyclical. I do not agree with some of the machinery which supports some of the ideas, that they spring from God, perhaps, I don't personally believe, or that some of these ideas are the natural consequence of ideas of earlier popes, in a natural and perfectly sensible way. I don't agree, and I will not ridicule it, and I won't argue it. I agree with the responsibilities and with the duties that the Pope represents as the responsibilities and the duties of people. And I recognize this encyclical as the beginning, possibly, of a new future where we forget, perhaps, about the theories of why we believe things as long as we ultimately in the end, as far as action is concerned, believe the same thing.

Thank you very much. I enjoyed myself.

Index

124
+

Index

About Richard Feynman

Born in 1918 in Brooklyn, Richard P. Feynman received his Ph.D. from Princeton in 1942. Despite his youth, he played an important part in the Manhattan Project at Los Alamos during World War II. Subsequently, he taught at Cornell and at the California Institute of Technology. In 1965 he received the Nobel Prize in Physics, along with Sin–Itero Tomanaga and Julian Schwinger, for his work in quantum electrodynamics.

Dr. Feynman won his Nobel Prize for successfully resolving problems with the theory of quantum electrodynamics. He also created a mathematical theory that accounts for the phenomenon of superfluidity in liquid helium. Thereafter, with Murray Gell–Mann, he did fundamental work in the area of weak interactions such as beta decay. In later years Feynman played a key role in the development of quark theory by putting forward his parton model of high energy proton collision processes.

Beyond these achievements, Dr. Feynman introduced basic new computational techniques and notations into physics—above all, the ubiquitous Feynman diagrams, which, perhaps more than any other formalism in recent scientific history, have changed the way in which basic physical processes are conceptualized and calculated.

Feynman was a remarkably effective educator. Of all his numerous awards, he was especially proud of the Oersted Medal for Teaching, which he won in 1972. *The Feynman Lectures on Physics*, originally published in 1963, were described by a reviewer in *Scientific American* as "tough, but nourishing and full of flavor. After 25 years it is *the* guide for teachers and for the best of beginning students." In order to increase the understanding of physics among the lay public, Dr. Feynman wrote *The Character of Physical Law* and *Q.E.D.: The Strange Theory of Light and Matter*. He also authored a number of advanced publications that have become classic references and textbooks for researchers and students.

Richard Feynman was a constructive public man. His work on the Challenger commission is well known, especially his famous demonstration of the susceptibility of the O-rings to cold, an elegant experiment, which required nothing more than a glass of ice water. Less well known were Dr. Feynman's efforts on the California State Curriculum Committee in the 1960s where he protested the mediocrity of textbooks.

A recital of Richard Feynman's myriad scientific and educational accomplishments cannot adequately

About Richard Feynman

capture the essence of the man. As any reader of even his most technical publications knows, Feynman's lively and multisided personality shines through all his work. Besides being a physicist, he was at various times a repairer of radios, a picker of locks, an artist, a dancer, a bongo player, and even a decipherer of Mayan hiero-glyphics. Perpetually curious about his world, he was an exemplary empiricist.

Richard Feynman died on February 15, 1988, in Los Angeles.